Building Wood Fires

THE COUNTRYMAN PRESS
A division of W. W. Norton & Company
Independent Publishers Since 1923

Building Wood Fires

ANNETTE
McGIVNEY

Techniques and Skills for
Stoking the Flames Both
Indoors and Out

THE COUNTRYMAN PRESS
www.countrymanpress.com
A division of W. W. Norton & Company, Inc.
500 Fifth Avenue, New York, NY 10110
www.wwnorton.com

978-1-68268-068-1 (pbk.)

10 9 8 7 6 5 4 3 2 1

Photo Credits

DEDICATION

For my son, Austin. You have been the brightest spark in my life
since the moment you were born.

CONTENTS

Preface

About five thousand years ago, a man hiked across the upper reaches of the Italian Alps. He was 45, had muscular legs to carry him over the mountain passes and was well equipped for the journey with a quiver of arrows, an axe, knife, and a complex fire-lighting kit that he wore on his belt. Unfortunately for the man, he was also gravely injured and he succumbed to his wounds high in the mountains where he soon became perfectly preserved in glacial ice.

After remaining undisturbed for five millennia, the mummified specimen of this Neolithic traveler—famously nicknamed Otzi, or the Iceman—was discovered in 1991 and proved to be a gold mine of information for scientists and anthropologists studying early human history. Everything about Otzi was fascinating to researchers but perhaps the most surprising finds came from his fire-making tools, which showed a technological sophistication that went well beyond what historians had thought existed during that era. Otzi's fire kit contained a dozen different plants for tinder as well as a highly flammable kind of fungus and flakes of pyrite for creating sparks. He also carried embers wrapped in maple leaves that were stored in a birch bark cylinder, what it took to make fire under any conditions. No matter how wet, cold, or windy it was in the mountains, he possessed the technology and skills to stay warm. And Otzi's discovery five thousand years after he went on his final journey reinforced to present-day anthropologists how entwined human evolution is with fire. Not only have humans been making fire since well before the Stone Age but we have been steadily perfecting its use to elevate our status on the planet.

"Utilizing fire is something we did that no other creature learned to do," says author and fire historian Stephen Pyne. "Fire was the earliest human technology and it fed everything else—tool making, ceramics, cooking. Even our understanding of chemistry comes from making fire."

But as ubiquitous as it may be to the human experience, making a fire quickly and successfully is not that easy. Or at least it is not easy for most of us who, unlike Otzi, have fallen out of practice with this basic skill. And that is where this book comes in. Perhaps it is because of my love of the outdoors but I have always been drawn to perfecting my fire-making skills. While friends of mine are drawn to the latest innovations in cell phone technology, I am more concerned with acquiring tools that can help me start a fire in the rain. The discovery of Otzi's fire kit was big news for me.

I want to share with you what I have learned over the years, not only about how to build fires in a variety of conditions but also how to appreciate fire in your daily life. Yes, fire is an important part of camping but it is also much more than that. This book will introduce you to many ideas for utilizing and enjoying a well-stoked blaze.

For starters, it helps to know what fire *is* exactly. This book begins with getting up to speed on the basic science of combustion as well as the mystical history of fire that inspired humans long before science existed. Then you will learn about primitive fire-making methods developed by Otzi and his ancestors as well as all the different modern methods that can be used for building the perfect campfire. Next, I hope to inspire you with ways you can incorporate fire into your home, both indoors and out. You will find here tips on how to turn your backyard into an "outdoor room" with a fire feature and how to efficiently heat your home with wood. You will also learn how to choose the best firewood and cook over coals.

In addition to sharing my own fire-making experience, I have sought out experts for this book whose lives revolve around fire. You will meet Stephen Pyne (quoted above), one of the world's most respected fire historians, as well as wilderness survival gurus, a national park fire-behavior specialist, urban and wildland fire fighters, fire ecologists, forestry experts, and a man who builds his own wood stoves. These pages are packed with practical advice on everything from how to buy the best woodstove to avoiding fire hazards to roasting a marshmallow. There are sidebars throughout that offer simple instructions on things like making your own wood carrier, how to build a fire-centric patio, and tricks for splitting firewood. And there is plenty of fun trivia here as well, including a scientific explanation on why we stare at a fire and also some tips on how to best utilize whiskey.

Ultimately, I hope you will find that once you get the hang of it, fire just feels good—whether it is on a campout, sitting around your backyard fire pit, or in front of a crackling blaze in your living room. So stoke those glowing coals, cozy up with this book, and enjoy one of the oldest pleasures known to humankind.

CHAPTER 1

Understanding Fire

The spirit and science of combustion

My relationship with fire has gone through various stages in my life depending on how much time I was spending camping and what kind of heating I had in my house. The period when I relied on fire the most was when I lived in a house where a single wood stove was the only source of heat. The house was located in the mountain town of Flagstaff, Arizona, at seven thousand feet with winter temperatures that were almost always below freezing. I was also raising my young son and I often felt like I spent more time tending to the wood stove to keep the fire going than I did caring for my child.

For the decade I lived there, I had a love/hate relationship with fire, as if it were an obstinate relative. Fire kept my family warm but it was so demanding. It lived and died, breathed and ate, was fickle and sometimes mysterious. I cursed the stove when I couldn't coax a fire to life and felt cared for by the cozy flames on frigid nights. Fire was like an ever-present spirit that inhabited my home just as it was for my ancestors thousands of years ago, and their ancestors thousands of years before that.

A Spark That Traveled around the World

For the first 4 billion years of the Earth's history there was no fire. There was not a sufficient level of oxygen to support combustion, nor were there sufficient carbon-based materials to supply fuel. The planet was plenty hot, though, from volcanic forces and the nuclear fusion of the sun after Earth spun away from its mother star. Some 400 million years ago, land plants finally grew plentiful enough to fuel fire but it was not until 120–200 million years ago that the Earth's atmosphere reached the oxygen level to allow for materials to consistently burn at a sustainable rate (i.e., catch fire but not be instantly consumed by the flames). When all the conditions were just right, a lightning strike likely lit the first wildfire, and fire has continued to burn somewhere on the planet ever since.

In his theories on the evolution of man, Charles Darwin assumed—as did many scientists who followed him in the nineteenth and twentieth centuries—that human control of fire did not come about until fairly recently when the modern humans first appeared in Africa about two hundred thousand years ago. Darwin noted that fire was "probably the greatest [discovery], excepting language, ever made by man." He figured that it would have taken a creature as clever as the human to figure out how to make and utilize fire. But that theory is changing. Some twenty-first century experts on early human history argue that it is the other way around. They believe fire use dates back to various ape-like species that existed long before the advent of modern humans. And these experts say it was fire, more than any other factor, that enabled the evolution of a single mammal over millions of years to become what we call a human.

Among the leading theorists in this new view of humans and fire is Harvard anthropologist Richard Wrangham, who points to archaeological evidence in Africa that ape-like creatures called habilines—considered the "missing link" between humans and apes—used crude tools like knives associated with cooking. Life was rough for the habilines, consumed by scavenging for food during the day, trying to avoid being eaten by the likes of saber-tooth cats at night, but suddenly—at least on the evolutionary time scale—some of the habilines began to evolve. The new creatures had significantly smaller jaws and bigger brains than the habilines; they could run and used spears to hunt large game; they had less hair and developed complex social systems. What could explain such a dramatic species shift? Wrangham argues that it was the harnessing of fire.

For perhaps 1 million years, these evolved habilines, called *Homo erectus*, harvested fire from the environment before knowing how to make it on their own. A member of a family band probably gathered a burning branch from a wildfire and carried embers from one location to the next, enabling the building of campfires for light and protection. According to Wrangham and others, the habilines who first harnessed fire learned that a campfire

YOU CAN TEACH AN APE NEW TRICKS

For more than two decades, American psychologist and primatologist Susan Savage-Rumbaugh conducted pioneering research into the communication skills of bonobos, a species of great ape. Her star research subject was a male bonobo named Kanzi who was born in captivity in 1980. Through Savage-Rumbaugh's training, Kanzi understands three thousand English spoken words. She also taught him to communicate using a keyboard that is labeled with geometric symbols representing familiar objects and activities. Kanzi uses the keyboard to communicate his desires in unprecedented ways and has mastered 348 symbols. One day Kanzi tapped on "fire" and "marshmallow." Savage-Rumbaugh obliged and gave Kanzi a pack of matches and a bag of marshmallows. She reported Kanzi then proceeded to snap twigs for a fire, lit them, and roasted a marshmallow over the flames. Although Savage-Rumbaugh has retired, Kanzi is still going strong at age 36.

Kanzi and his bonobo family live in a $10 million 18 room house and laboratory complex built for them at the Iowa Primate Learning Sanctuary outside Des Moines. According to a 2008 article in *Smithsonian* magazine, Kanzi and his clan spend evenings sprawled on the house floor snacking on blueberries, M&Ms, and celery. They select DVDs they want to watch by pressing buttons on a computer screen. Their favorite movies are *Quest for Fire*, *Every Which Way but Loose*, *Greystoke: The Legend of Tarzan*, and *Babe*.

scared away predators, which allowed the creatures to sleep on the ground and in caves instead of in the trees. It lengthened waking hours, which allowed for a host of new nighttime activities: tool making, tending to the injured, and socializing, perhaps leading to the foundation of language. The growing social structure and increasing ingenuity facilitated hunting and, eventually, cooking.

Cooked meat is more nutritious and easier to chew and digest. Consuming it regularly led to an increasing brain size with a larger skull and a smaller jaw. Instead of lots of hair, it was animal hides and sitting next to the fire that provided warmth—enough warmth to enable *Homo erectus* to migrate into colder climates, even during glacial periods. Cooking over the fire also promoted the first kind of domesticity because the meat was taken back to a protected home base and shared among the group rather than eaten immediately in the field. "We humans are the cooking apes," writes Wrangham, "the creatures of the flame."

Learning how to start a fire most likely came about accidentally as a result of tool making. The rubbing of wood and striking on rocks for tools produced smoke and sparks. It was probably not a big leap for *Homo erectus*, with his increasing brain size, to make the connection that this was the same kind of fire he had been harvesting from the forest. And he began experimenting and innovating to develop one of the most foundational technologies of humankind. Considering that there were glacial periods covering parts of the Earth during the one million years *Homo erectus* roamed the planet, it seems unlikely the species could have survived migrating into colder, potentially ice- and snow-covered landscapes without knowing how to make fire. Consequently, when *Homo sapiens* first came onto the scene two hundred thousand years ago they were already equipped with ample knowledge of not only how to make fire but how to structure a successful life around its myriad uses.

Eternal Flame

Given the reliance of early humans on fire for survival, it is no surprise that creation stories for many indigenous cultures are centered in tales of discovery and mastery over combustion. Plus, as I discovered during the taxing wood-stove period at my mountain house, fire has a life of its own. Unlike other tools used by early humans to survive, such as stone knives, clothing, or shelter, fire is alive and sometimes it seems to be calling the shots. During most of human history, attaching super natural qualities to fire was the only logical way to explain it.

According to Scottish anthropologist James Frazer, nearly every culture on every continent,

from Tasmania to South America to ancient Greece, attributed their existence to fire. Some Indonesian peoples "say that the Creator made the first man and woman by carving figures in human shape out of stone and causing the wind to blow on them, for thus they acquired breath and life," wrote Frazer in 1930. "He also gave them fire, but did not teach them how to make it. So in those early days people were very careful not to let the fire go out on the hearth." The story continues to recount how fire was lost when it was accidentally allowed to go out and the people had to trick the gods to learn how to make fire on their own. In this case it was by "striking a flint with a chopping knife."

Likewise, one local story from Paraguay recorded by Frazer describes how the indigenous Lengua people learned to cook with fire. A man who had been out hunting came upon a bird that was cooking snails over a fire. "There he observed a number of sticks, set point to point, the ends quite red and giving forth heat. . . . Being hungry, he tasted the cooked snails, and finding them delicious he made up his mind that he would never eat raw snails again." The Utes of southwestern Colorado told a legend about getting fire back after it had been stolen. "Coyote caught fire and gave it to the Indians," wrote Smithsonian anthropologist Walter Hough in 1929. "The Indians kept the fire and never lost it again. It made light and heat. It was cold, and if there had been no fire the Indians would all have died."

Fire was also central to the mythology of ancient Greece and Rome. The Greek god of fire was Haitos

Festival of Saint John's Eve

and his counterpart who presided over the home was Hestia, the goddess of the hearth. All Greek cities had a temple in the center called a prytaneum where the "sacred city fire" was kept continuously burning. When a new territory was established, the Greek colonizers carried fire from the local prytaneum so they could use the flames to start a perpetual fire in the next temple. The perpetually burning torch of the Olympic games still in use today is a continuation of this tradition. Rome also had a temple in the middle of its cities containing a fire that was kept burning continuously and was dedicated to Vesta, the goddess of the hearth. The temple fire was tended by four to six young women who came from Rome's leading families. These "Vestal Virgins" held the prestigious post for a thirty-year term before they were released back into the community.

In Judeo-Christian tradition, God often communicated with humans through fire. In the book of Exodus in the Bible, Moses receives divine instructions on leading the Israelites out of Egypt from God in the form of a burning bush. In the New Testament, the Holy Spirit first appeared to the disciples of Jesus in the form of a fire above their heads.

Fire continued to be part of spiritual rituals across the world well into the Middle Ages, including in Europe with the burning of large bonfires on June 23 for the Festival of Saint John's Eve. Connected to the summer solstice, the fires were believed to protect against evil spirits who roamed freely when the sun turned southward. At Christmastime, the end of an entire tree was burned in the family hearth to mark the highlight of winter solstice celebrations, called a Yule log.

However, by the mid-eighteenth century, the Age of Enlightenment and the emergence of science was displacing fire mythology with newly discovered facts. And while Vestal Virgins were no longer needed, fire continued to play a central role in human evolution. Only now fire became a means for understanding emerging scientific principles. Experiments involving the burning of a wide variety of materials by alchemists and nascent scientists helped bring a new understanding about how the natural world worked. Nearly a thousand years earlier, the leading Greek thinker Aristotle had introduced the long-held view that fire, along with water, earth, and air were the four basic natural forces that made the world go round. But the more scientists in the 1700s played with fire, the more they realized those forces were something else—atoms and molecules, and energy from the chemical reactions between them.

"Compared to the other three elements, which are substances, fire is a reaction," explains fire historian and author Stephen Pyne. "Fire is alive. It moves. It synthesizes its surroundings." This new understanding of fire became a foundation of modern physics, chemistry, and biology. It also led to major technological innovations, including the internal combustion engine, which propelled the Industrial Revolution into the twentieth century.

The Physics of Fire

What eighteenth century scientists figured out—and what children have learned in school ever since—is that fire, or combustion, is essentially a chemical reaction between carbon and oxygen. Carbon is the most abundant chemical element in living things (trees are about 50 percent carbon) and the bonds between carbon atoms store a large amount of chemical energy. When carbon bonds break in the presence of oxygen, they release energy in the form of heat and light.

"You can explain fire in terms of physical chemistry. It is the oxidation of hydrocarbons. But I like to think of it as a product of the living world because life created the oxygen and life created the fuels," notes Pyne. While the "oxidation of hydrocarbons" may sound complicated it is the same process that the human digestive system uses to metabolize food, albeit at a different rate. A person's digestive enzymes slowly disassemble the molecules in plants by breaking the carbon bonds so that the energy stored in the bonds can be used to nourish the body.

In the late seventeenth century, alchemists in Europe experimented with the first versions of matches and frequently had disastrous results. These early inventions were dipped in white phosphorous and often would spontaneously combust when removed from a box or even inside someone's pocket. A later variety combined sulphur and phosphorous to make the match more stable but the matches produced showers of sparks and choking fumes of sulphur dioxide.

The first match suitable for widespread use, called the safety match, was patented in 1852 in Sweden. The tip of the match was dipped in red phosphorous, which is more stable and less toxic than the white variety. But users still had to be careful to keep the match head and the striking strip separate to avoid accidental combustion. The "strike anywhere" match was patented in 1898 in France and contained a tip made from the relatively harmless sulphide of phosphorous instead of the dangerous white form.

Present day matches have changed little from those first developed in the nineteenth century. The tip is a mix of red phosphorous and potassium chlorate. When the chemicals heat up through the friction of striking, they mix to produce ignition and then a flame. The wooden match stem usually has a fire retardant on it to allow it to burn more slowly.

You can make your own waterproof matches simply by taking strike-anywhere matches and dipping them in a thin coating of melted beeswax. Allow the matches to dry on a piece of foil and then carry them in a sturdy container such as a pill bottle or Altoids box. The wax will sustain combustion even when wet.

Meanwhile, in the process of combustion, the stored energy of organic material (i.e., wood) is released instantaneously by fire when oxygen and heat combine to break the material down.

The "fire triangle" presented in most grade school science classes offers the simplest model for the three requirements of combustion: oxygen, heat, and fuel (a.k.a. carbon). And there must be appropriate amounts of each. If any side of the triangle is missing, a fire cannot start. If any side of the triangle is removed the fire goes out. Based on its flaming appearance, it might seem that the sun is on fire. But there is no oxygen or fuel on the sun to support combustion so there is no fire. It is just extremely hot and bright due to the same kind of nuclear fusion that produces a powerful reaction in a hydrogen bomb. Even molten lava from a volcano is not technically on fire until it comes in contact with some kind of fuel.

For the purposes of building campfires, getting access to air and fuel seems easy enough on Earth. Carbon and oxygen is everywhere. But heat? It is a reaction that only occurs when molecules get excited and start moving around. The faster the molecules move the hotter they get, just as when you rub your hands together to warm them up. This is why the easiest way for humans to start a fire is to create an ignition source through friction, whether it is rubbing two pieces of wood together or striking a match.

Anyone who has lit a match and thrown it on a pile of tinder only to watch the match burn down and never ignite the tinder has experienced a frustrating reality of physics: a flame is not really a fire. It is simply a source of ignition to generate heat that could lead to combustion of fuel materials if things are done right. Or if the fuel is not properly assembled (see Chapter 3), the flame will just die out. That's because a flame is simply the visible chemical reaction of vapors rising into the air from a fuel source agitated by heat. The flash point for wood, which is the temperature necessary for the substance to ignite into a flame, is 572°F (although this can vary somewhat depending on ambient temperatures and humidity levels away from a laboratory setting). But the flash point for the phosphorous compound on the tip of a match is much lower at 375°F.

If the heat from a flame successfully transfers to the surface area of fuel, then there is fire.

From a molecular standpoint, there are three ways heat energy can transfer from one object to another: conduction, radiation, and convection. In conduction, the heat energy passes through other molecules, before reaching its combustion destination, such as when heat from a stove travels through a metal kettle to heat water for tea. In radiation, the heat travels through the air to heat an object, such as hands being warmed over a campfire. Convection is direct heat transfer, with the excited energy molecules in one object heating up the molecules in the object next to it. While radiation is often a secondary source of heat transfer in a wood burning fire, the primary combustion process is fed by convective heat. Combustion happens when sustained heat—usually by convection—increases the temperature of the fuel to the point that the molecules begin to break down as they combine with air, just as digestive enzymes break down food in the human body.

However, appreciating the physics of a single flame provides a foundation for not only understanding fire but also the basic principles of nature. In the mid-nineteenth century when the Western world was becoming fascinated with science, Michael Faraday, a scientist at the Royal Institution in London, gave

a series of popular lectures in which he said: "There is no more open door by which you can enter into the study of philosophy than by considering the physical phenomena of the candle. There is not a law under which any part of this universe is governed which does not come into play, and is not touched upon, in these phenomena."

Faraday explained to packed auditoriums how the newly discovered principles of physics and chemistry were demonstrated in something as simple and ubiquitous as a candle flame. He demonstrated how a candle flame is a colorful, vaporous bag that rises above a pool of melted wax. The outer wall of the bag is blue because this is the area where the chemical reaction of oxidation is taking place as the hydrocarbon-based molecules in the wax mix with the oxygen molecules in the air. Fuel in the form of liquid wax travels up the candle's wick and releases carbon-based soot particles into the flame's vaporous cloud. As the soot particles get hotter they change in color from red to orange, then yellow, and finally blue. But if combustion is incomplete they cool and turn into black smoke. Molecules that escape the flame travel into the air as water vapor and carbon dioxide. The hottest part of the flame is in the blue combustion reaction zone and the coolest is the black tip of the wick at the bottom of the flame. The light produced by the candle flame is from the yellow and orange incandescent soot particles.

Faraday entertained the crowd by conducting

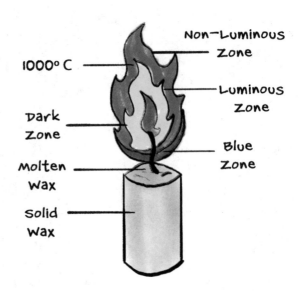

experiments to show ways heat and vapor could be adjusted to produce various reactions. He explained how the chemical process of a burning candle remained perfectly steady and maintained an oxidation rate that allowed the flame to stay relatively the same size as the candle burned down. It was this new scientific understanding of steady energy output and the incandescent soot particles that soon led to the invention of the incandescent light bulb by Thomas Edison in 1879.

But eventually the candle burns down or the match burns out. Knowing what makes a flame transform into sustained combustion is essential for anyone desperately trying to start a fire on a cold night.

Flame to Fire

As described above, once the flame of ignition is transferred to a fuel source, the surface of the fuel needs to adequately heat up in order for it to catch fire. This convective heat process initially happens on the surface of the fuel, so the greater the surface to volume ratio of the fuel, the faster it becomes hot enough to ignite. This is why pine needles or paper catch fire more easily than a giant log.

Another factor impacting combustion is moisture. When a flame first touches fuel, a large amount of the heat energy is spent drying out the material. Water is a much better conductor of heat than carbon-based materials like wood, so it sucks up the ignition flame. This drying process produces smoke comprised of water vapor and soot that is released into the air during incomplete combustion. If the fuel is too moist, combustion never happens because the wood cannot get to that magic flashpoint of 572°F.

However, when burning wood, keep in mind that smoke is the first sign that your fire building is moving in the right direction. Once wood is heated up to about 300°F not only is it drying out the water in the wood but the heat is also releasing hydrocarbons such as sap. Between roughly 300–572°F, as the heat evaporates the water and hydrocarbons, the vapor rises in the form of smoke. For the sake of comparison, charcoal briquettes on a grill catch fire without producing any smoke because the briquettes are pure carbon with all vaporous materials already removed.

But with wood, heat started by ignition of tinder begins to break down the carbon molecules. Once the heat finally reaches the flash point, the process progresses from smoke to complete combustion, often with a popping sound in the fuel. Now the molecules are released as a cloud of vaporized fuel that is engulfed in flame—the same kind of flame that is produced by a burning candle. However, wood does not burn nearly as cleanly as candle wax, so it also produces soot and smoke, although far less than when the fire was first starting. As more tinder catches fire and the fuel cloud becomes larger, the increasing flames transfer more and more heat onto the surface of the logs.

Once the heat becomes great enough that the flames are no longer just charring the exterior of the logs but causing the outer surface to break down and release fuel vapors of their own, then combustion is self-sustaining. It is time to sit back and enjoy the fire.

During the combustion phase, radiant heat emitted from burning logs dries out and heats up unburned logs, priming them for ignition. While combustion of wood first happens in a flurry of flames, as the process progresses and the burning wood breaks down, the log moves into a glowing stage where less gaseous fuel is being released but the oxidation process is going strong. These steadily burning, glowing coals give off more heat than a

THE UPS AND DOWNS OF THE CARBON CYCLE

Smoke from wood burning fires has received a bad rap in the last several decades as problems with air pollution and concerns over climate change are of increasing public concern. But what many people don't know is that fire—and smoke—from burning trees is part of the Earth's natural, life-sustaining carbon cycle.

Forests, including the trees and the soil beneath them, store carbon. Eventually, the carbon is released into the atmosphere, either rapidly through burning or slowly through decomposition. The carbon combines with oxygen in the atmosphere to form carbon dioxide (CO_2), which is then absorbed by new plants through the process of photosynthesis. Without the carbon, the new plants could not grow and thrive. Essentially, the carbon released from dead or burned trees sustains the new trees that replace their predecessors.

Because of this, some ecologists and other scientists argue that burning wood for heat is a carbon neutral form of energy because the carbon in the trees has to be released one way or another. This contrasts with fossil fuels, which are not part of the planet's active carbon cycle because the CO_2 from petroleum, coal, and natural gas has been sequestered deep in the Earth until it is added to the atmosphere through burning.

However, the carbon neutral theory on deriving energy from wood burning only holds up under certain conditions. The wood must come from limited amounts of fuel harvested from sustainably managed forests. And then it should be burned in moderation—small caveman fires, not industrial infernos. Also, if a large swath of forest is denuded as the result of a catastrophic wildfire, far more carbon is sent up into the atmosphere at one time than new plants, once they finally start to grow, can consume in the years following the fire. It's also possible that the Earth's warming temperatures due to climate change could make it difficult or impossible for the same native trees to take root and thrive after a massive fire because of invasive species that are more easily adapted to the changed landscape.

Yet, when practiced in a balanced way, harvesting and burning wood helps maintain the Earth's natural carbon cycle. Life on Earth was designed to depend on fire.

flaming log and are ideal for cooking or marshmallow roasting. Eventually, the burning logs reach the smoldering stage, the final phase of combustion. Wood that does not decompose to ash remains behind as bits of char that is almost pure carbon.

Other factors that impact the combustion process are the size and shape of the logs. Generally, there is simply not enough sustained heat during the early stage of a fire to break down big logs. And while denser or harder woods will burn longer they are tougher to get going. So, hardwoods and the big logs should be placed on the fire only after you have several logs in the glowing stage of combustion to ensure sufficient heat transfer to the unburned hardwood.

As Stephen Pyne sagely put it, "fire is a relationship." During the early days of my living in the house with the wood stove, that relationship was pretty rocky. One night in a frustrated fit, I threw expensive tequila on some combustion resistant logs to try and jumpstart a fire with alcohol. It didn't work. Eventually I learned the art of tending and feeding, moving glowing logs around and interspersing them with unburned pieces of fuel. And even if this process could all be explained by science there still seemed to be a bit of magic required.

FIRE KEEPER: DANIEL PEARSON

Daniel Pearson grew up in a family where his father and cousins were volunteer fire fighters and he wanted to be one, too.

"Like most people I saw fire as something that was dangerous when it got out of control," says Pearson. "It was something to be feared, to be fought."

After working as a structure firefighter, Pearson moved to wildland fire fighting on public lands. And the more he witnessed fire in its natural setting, the more he fell in love with it. "I came to see fire as being alive and beautiful," says Pearson. "Fire is a natural agent of change and essential to a healthy ecosystem. If we don't allow fire to happen, environmental problems continue and we are just kicking the can down the road."

Pearson went from being a wildland fire fighter to a fire behavior analyst. He has worked at Grand Canyon National Park for the past 12 years and is part of a team of scientists who use controlled burns to improve the ecology of one of the world's greatest natural wonders. But figuring out the behavior of a fire is tricky business. Pearson plans prescribed burns years in advance to mitigate environmental and social impacts. And then he is constantly adjusting his plans based on prevailing winds, humidity levels and ambient temperature—all of which impact fire behavior. He tries to carry out his work as invisibly as possible. When park visitors have traveled half way around the world to see Grand Canyon they don't want to find it filled with smoke.

Thanks to the work of Pearson and others on the park's fire management team, the ponderosa pine forest on the rim of Grand Canyon is more open and healthy than it has been in decades. And there are other perks. Pearson enjoys the panoramic view that comes with his job but he also enjoys the aroma.

"If we are burning piles [of trees], I'll hang my shirt next to them to soak up the aroma," he says. "I adore the smell of smoke from a natural fire."

CHAPTER 2

First Fire

The most primitive methods of harnessing flames

Once you understand what fire is and how it works, it's worth understanding the earliest methods for controlling it. Like most Westerners in the developed world, I didn't think learning to ignite a fire using ancient techniques mattered anymore. The idea of teaching myself how to start a fire the way cavemen did, basically by rubbing two sticks together, actually seemed a little silly and

also probably impossible. I thought the manual dexterity, stamina, and skill required to perform this obsolete task must have surely been erased from my evolutionary DNA thousands of years ago. It would be like asking a caveman to tap out a text message on a cell phone. But since I wanted this book to be as comprehensive as possible, I agreed to give it a try.

I visited retired Army officer Al Cornell at his home in the high desert of northern Arizona.

Cornell, 74, is something of a primitive fire fanatic. His garage is filled with fire-making tools, the same kind used by humans ten thousand years ago and earlier. Cornell first learned primitive fire-making methods in 1967 when he was stationed in the jungles of Central America. Since then, he has collected ancient fire-making implements from around the world and taught himself how to use them. Cornell passes the skills on by offering fire-making demonstrations to Boy Scouts, hiking clubs, Native

American tribes, and anyone who is interested. Now that included me.

Cornell's enthusiasm for the lost art is contagious. He suggested that I try my hand at the bow drill. A bow drill set was once ubiquitous in the human home, probably hanging on the wall in huts and caves around the world. Archaeologists found one of the earliest-known bow drills in Egypt that dated to approximately six thousand years old. Bow drills have also been discovered at ancient Native American sites in the Southwest that date to 1100 AD. This primitive technology involves the bow, which is made of a thin, sturdy branch that has a cord attached on each end. The cord wraps around a hard, straight piece of wood called the drill or spindle that is held perpendicular to the bow. A round socket is held in one hand and positioned on top of the drill to keep it in place while the user spins the drill with a back and forth motion of the bow. The bottom end of the drill is carved into a tip and positioned in a notch on a piece of flat wood called the hearth. With the right amount of downward pressure on the drill, and back and forth motion of the bow, friction will produce wood dust, then smoke and, finally, an ember at the tip of the drill.

Cornell makes his bow drills from natural materials he collects on public lands around his house.

Fire being started using a bow drill

Being completely authentic and historically correct, both in terms of materials and practice, is of supreme importance to Cornell. For my demonstration, Cornell knelt on one knee on his driveway with the hearth board pinned against the pavement beneath his foot. Using an elk bone for the socket, he kept the drill in a steady position with one hand while he pulled on the bow with the other. In little more than a second there was smoke, and in maybe another three seconds there was the spark of an ember. He dumped the ember into a bundle of dried juniper bark and gave one long exhale. The bark exploded into flame.

"Now it's your turn," he said, handing me the bow and drill.

While Cornell made this look as simple as whipping up a peanut butter and jelly sandwich, I seriously doubted I could do it. I confessed that I had actually tried to use a bow drill during a two-day outdoor survival course years ago and found myself impossibly incompetent—as did most of the other one dozen students on the course. I was positive this time would be the same.

"Well, just give it a try," encouraged Cornell. "I think you are going to be surprised."

He positioned my foot on top of the hearth board and my hand on the elk bone, telling me to press down gently on the drill. With my other hand I slowly moved the bow back and forth, holding it parallel to the ground, getting a feel for how to make the drill spin without losing control and causing all the pieces to scatter across the driveway.

"Now, make longer strokes while picking up the speed," advised Cornell.

I tried to go faster but my strokes were clumsy and uneven. There were so many moving parts to keep in place. I started to sweat.

"Faster!" said Cornell, seeing that I was losing steam just as a few feathers of smoke emanated from the bottom of the drill. Then the smoke disappeared. I couldn't do it.

"Let's try a different drill," suggested Cornell. "I think you might do better with willow."

It turns out that customizing technology to fit personal preferences goes way back. Cornell suspected my body mechanics were not suited to the cottonwood drill I was using. I resumed the position on the pavement again, but this time with the elk bone pressing down on a piece of willow the size of a fat Cuban cigar. And something flipped inside me as I pulled the bow back and forth. I increased the speed of the strokes. I had the downward pressure on the drill more dialed in. Soon there was smoke.

"Keep going! Enjoy the ride!" encouraged Cornell. He sensed a shift in me. Perhaps he would win over another convert.

"Look at that!" he exclaimed. "You've got an ember!"

He took the hearth board from under my foot, tipped it into another tangle of juniper bark and handed the smoking bundle to me.

I spewed air onto the bark like I was trying to

WHY DO WE STARE AT THE FIRE?

Regardless of how hectic the day may have been and how frantic things seemed when setting up camp, within five minutes of sitting around a well-stoked fire, the anxiety level seems to drop to zero. Whether it is enjoyed in a campground or in a home fireplace, there is no denying that fire has a powerful calming effect on people. Most other creatures are afraid of it, but humans are drawn to fire. Enjoying the emotional and psychological comforts provided by an open flame is one of the main reasons why people go camping. And some people crave the fire fix so much at home that if they live somewhere without a fireplace they stream a video of a fire onto their television.

Considering the positive relationship that humans have had with fire for more than 1 million years, it is no mystery why a campfire brings comfort. There is a message stamped in our DNA that tells us we are safe and our odds of survival are good as long as we are near a campfire. But what is it about a fire that makes humans want to stare at it for hours on end, as if in a hypnotized trance?

There has been little research into this topic but anthropologist Frances Burton has a potential answer. Burton researched how firelight at night may have impacted the development of *Homo erectus* and early pre-human primates. She theorizes that sitting around a campfire millions of years ago not only had physiological effects but also psychological ones in the way it shifted the species' relationship to night.

Burton's research involved measuring the hormones of human participants who stared at a campfire for an extended period. She found that the firelight significantly reduced the secretion of melatonin, the hormone that induces sleep. And with the decrease in melatonin came an increase or a spike in dopamine, the neurotransmitter in the brain that promotes good feelings. You can also get a spike in dopamine from biting into a bar of delicious chocolate or getting a kiss from someone you love. But staring at a fire is perhaps one of the oldest ways humans have been able to get that feel-good fix.

Burton also found that simply sitting around the fire is not enough. Her research showed that participants who kept a fixed gaze on the fire experienced the greatest effects. In fact, she reported that staring at the blue light at the bottom of the campfire—the hottest part—reduced melatonin levels the most. In a separate test, Burton had subjects stare at a green light emitted from a bulb and there was no similar fluctuation in melatonin. As Burton wrote, fire made "cold seasons warmer and nights shorter, in effect, creating microclimates amenable to the Ancestor's way of life."

blow out fifty birthday candles. Poof! It exploded into flame and I placed the burning bundle in a metal pan.

"My first fire!" I shouted jubilantly. I felt as if I was five and had just learned how to ride a bike. I was not only pleased that I was able to successfully start a fire with the bow drill but also surprised at how satisfying it was. I had started plenty of camp-fires during my life with matches and lighters but this was different. There was something deep down that was kindled, as if I had tapped into my inner cave woman. I suddenly understood Cornell's fasci-nation with this forgotten art.

Not only has the skill of making fire with prim-itive methods been relegated in our culture to the stuff of museums, but fire itself has been generally eliminated from our daily lives. Even though our homes, cars, and cell phones still rely on combustion to function, the fire that powers these appliances, vehicles, and computers happens out of view and often far away. But that does not change the fact that our relationship with fire is deeply embedded in our DNA. It is who we are as humans. Understanding fire means understanding ourselves as individuals with an intrinsic connection to the natural world.

Making Fire and Other Pyrotechnics

Depending on the organic materials available and environmental conditions in a geographic region, early methods for starting a fire varied but they gen-erally fell into two basic categories: rubbing wood together to create friction or striking mineral rock to produce a spark. And just as electronics have expo-nentially improved over a brief time period in the twenty-first century, early humans were continually improving fire-making technology by perfecting methods and trading materials with other groups.

Universal fire starting methods include:

Hand drill: This is perhaps the simplest and old-est method practiced by indigenous cultures around the world. It involves rubbing hands up and down a wooden spindle that is positioned in a depression on a hearth board. The faster the hands move, the more the drill spins, until enough friction produces an ember on the hearth. Although it may appear simple, successfully getting an ember requires a great deal of experience as well as physical effort, not to mention some heavy calluses on the hands. Generally, Native American cultures throughout North America used a hand drill in prehistoric times made from hardwoods such as cottonwood, willow, greasewood, and cedar. As with all wood-friction methods, it is important that the drill material is slightly harder than the hearth material in order to create wood dust that would turn into an ember.

Primitive fire expert Al Cornell says the acqui-sition of the right materials was something that took a great deal of thought and expertise: "Our hunter-gatherer grandparents over thousands of genera-tions did not simply walk into a campsite after an exhaustive day of hunting and randomly grab two nearby sticks from the forest and start a fire. They

were actually starting their fire-making process many hours before when they were collecting materials as part of their gathering activities, or even days before when they put together tested materials into fire kits in anticipation of this future need."

Bow drill: The bow drill was the fire technology equivalent of the iPhone 2, marking a significant improvement on the hand drill because this next generation required less physical effort and usually produced an ember quicker thanks to the mechanical automation provided by the bow. It was possibly developed in Egypt and also used by the Greeks and Romans, as well as in Mesoamerica and by ancestral Puebloan cultures in the Southwest. See the introduction to this chapter for a fuller explanation of how it works.

Strap drill: This method was developed by Inuit ancestors and was popular among Arctic and Sub-Arctic peoples. The regional appeal was probably due to the fact that the materials were small, portable, and easily carried across snow-covered areas where wood for fire-tool making was limited. The strap drill was considered so important to the Inuit that when a person died, his strap drill was placed in the grave with him.

Similar to the bow drill, there is a socket in which the drill spindle is positioned. But the socket, which was usually made from a piece of bone, can be held in the user's mouth between the molars. The other end of the spindle goes in the hearth board and a cord wrapped around the spindle is held on both ends and positioned parallel to the ground.

With the socket in your teeth while kneeling over the hearth, you rapidly pulled the cord back and forth to create friction of the spindle tip on the hearth.

After my success with the bow drill, Cornell convinced me to give the strap drill a try. But I could not get an ember—not only because I lacked the proper finesse in pulling on the cord but also because a steady flow of my drool kept dousing the heat on the hearth.

Saw or plow: This wood-friction method was widely used in regions such as southeast Asia where there was a plentiful supply of smooth wood like bamboo. For the plow, rub a stick with a sharpened tip rapidly back and forth in a smooth channel running down the center of a hearth board. For the saw, place the stick in a slot on the hearth running perpendicular to the length of the board, then rub it back and forth to generate heat and an ember.

Pyrite and flint: As the dominant method used by ancient humans across the European continent, this technology was commonly called "strike a light." It was completely dependent on having some sort of hard iron sulfide rock, usually in the form of marcasite or pyrite. The marcasite or pyrite could be struck against another piece of the same material to throw a spark or it could be struck against flint, which is made of hard quartzite.

This was the most widely used fire starter from Paleolithic times through the Iron Age, around 1200 BC, when pyrite was replaced by steel as the

TRY THIS: MAKE YOUR OWN HAND DRILL

Even though the hand drill appears to be the simplest of the wood-friction fire-starting methods, it is the most difficult in terms of the skill and physical effort required. But mastery is possible with practice and a fair amount of sweat equity. Here is what you need to get started:

Hearth board: This should be a piece of relatively hard wood, such as aspen, willow, cottonwood, or sycamore. The harder woods such as hickory or cherry can actually be too hard for drilling. However, soft resinous woods such as pine or spruce are too soft for the purposes of creating friction and grind down too quickly. If you are working with a log or branch, split the wood into a piece that is flat and about 1/2 to 1/4 inch thick. The length is not as important but it should be long enough for you to stabilize with your foot and still have room for drilling—so about 8–10 inches. The width of the board doesn't matter as long as it is wider than the diameter of the drill.

Drill: The most important aspect of this technology is that the drill must be made from a very straight piece of wood in order to work. And it also has to be round so you can spin it in the palms of your hands. The diameter should be about the size of your index finger. It also needs to be strong and at least as hard as the hearth board, preferably a little bit harder. The length should be about 2½ feet. The wood also should be dry (not green and pliable). Aspen and willow shoots can be retrieved and then straitened while drying over a period of several weeks. Other good options include the stalks of large flowering plants such as yucca, mullein, and sunflowers. Use a knife to shave the drill tip into a dull octagon-shaped point.

Technique: Getting an ember with a hand drill has to do with achieving just the right combination of spinning speed and downward pressure on the drill. But first you need to create the hearth, which is the shallow depression that will become the receptacle for the friction-generated heat. Create the hearth first before you start drilling in earnest. Holding the hearth board steady under your foot or knee place the drill tip about 1/16 of an inch from the edge of the board. Spin the drill until you create a slight depression. Then take a knife and cut a v-shaped notch from the rim of the hearth circle down the edge of the board, almost cutting through to the bottom but not quite. This notch will collect your hot wood dust and eventually the ember.

Next, put the drill tip back in the hearth and start spinning, moving your hands rapidly back and forth, going from the top of the drill to about six inches above the board. Keep repeating this motion, moving your hands quickly back to the top of the drill while also maintaining a steady downward pressure. Make sure you never remove both hands from the drill as you are moving back to the top or you will lose momentum and pressure of the drill bit on the board. Keep the drill fluid in between your palms as you run your hands up and down, almost rhythmically as if you are playing a musical instrument. If you get winded, take a break and then start again. It is more effective to drill in short energetic bursts.

FIRE KEEPER: AL CORNELL

When Al Cornell retired from the Army in 1996, he began indulging his fascination with primitive survival skills. He took classes in animal tracking and began making Stone Age tools and arrowheads out of flint. During a tracking course with survival guru Tom Brown, Cornell was introduced to the bow drill and hand drill fire methods. "All of a sudden, all I wanted to do after that was make fire," recalls Cornell. "It was like some imprinted DNA was awakened in me and fire was more fun than anything else."

Cornell taught himself how to use the hand drill and began a ritual of starting a hand drill fire every three or four days to stay in practice. He marks each fire on a calendar and the thick calluses on his hands attest to how many times he's done it. In November 2016, Cornell reached the benchmark of eight thousand hand drill fires. He has also devoted much of his retirement to researching primitive fire methods and the cultural histories behind them. Cornell traveled to Alaska to investigate the Arctic strap drill and went to Panama to meet indigenous cultures using the bamboo plow. He has shared his knowledge during demonstrations at the Smithsonian in Washington D.C. as well as with Native American tribes who wanted to learn lost traditions.

When information about primitive fire methods is lacking in university libraries, Cornell has conducted field experiments to try and fill in the knowledge gaps about how early fire technologies were implemented. He has experimented with different types of wood for hand drills, explored how humidity levels impact fire starting in the jungle, and investigated types of fungi that were likely used as tinder with pyrite. All of his experiments are carried out with naturally harvested materials to authentically replicate the conditions experienced by ancient humans. In order to evaluate the benefits of cooking that fire provided, Cornell even ate raw meat to see how much longer it took to chew than the roasted variety. The result: An ounce of raw elk meat took nine minutes to chew compared to five minutes for a cooked ounce of the same meat.

Cornell frequently goes hiking around his home and he likes to collect fire-making materials as he goes, just the way the earliest humans did. When he stops for lunch he usually makes a small fire using a hand drill (either manufactured on the spot or brought along) and toasts his sandwich on a small grill.

"Fire is never boring to me," says Cornell. "Practicing these skills is a very real way to connect with our prehistoric past."

starting material. However, throwing a spark is only half the battle with this method, especially in the soggy conditions that were typical of the northern European climate. Various species of fungi were collected as tinder from the forest and ground into powder. A favorite was *Fomes Fomentarius* which would reliably transform a spark into an ember and burn for several minutes even when wet.

CHAPTER 3

Campfires 101

The perfect campfire is possible with the right kind of skills and materials

---

When it comes to fire starting today, the wood-friction methods are difficult at best to successfully perform, and because of the practice required, they are not a prudent option for campers in a survival situation. Like most people, wilderness survival instructor Tony Nester usually starts his fires with matches or lighters. But he relishes the special times when he starts a fire the ancient way. I came to appreciate this as well when I finally learned to operate a bow drill thanks to Al Cornell's coaching. It is not necessarily a practical skill but it is a magical one.

"For me, starting a fire with a lighter does not produce the same satisfaction that I get from a fire I started with a hand drill from materials that I gathered from the surrounding forest," says Nester. "After I have carved the drill and put my sweat into getting the fire going, I feel like I am part of something bigger. I am integrated with the surrounding environment and all the other elements beyond my own actions that combined in that place at that time so I could be warm."

Tony Nester is the last person you would expect to have trouble starting a campfire. He regularly teaches U.S. military Special Operations units how to survive in the wilderness with little more than a knife, and he can coax flames out of thin air simply by rubbing two sticks together just as early humans did. But Nester likes to tell his survival school students about a family camping trip he took not long ago in Colorado when he did everything wrong.

"It had been a long day of driving when we finally pulled into the Forest Service campground just as it was getting dark," recalls Nester. "We were hungry and wanted to cook dinner. We still had to pitch the tent. And I was in a hurry to get the fire going." Nester placed several big logs he brought from home in the campsite's fire ring. Then he stuffed the Sunday newspaper and some tree bark underneath the logs. He put a match to the newspaper and it caught fire within seconds as Nester turned his attention to other camp chores. Then the flame went out once the paper and bark had burned up completely. Nester's tired, hungry children looked on with puzzled disappointment.

But it was no mystery to Nester why the fire failed to ignite. "I skipped the middle steps," he says. "I was in a hurry and just throwing things in haphazardly and not taking the time to add more tinder and kindling."

In his outdoor skills classes Nester teaches a four-step process to reliably make a flame transform into fire. "It doesn't matter who you are and what your experience level is in the outdoors," he says. "Getting a fire going comes down to basic science.

If you don't follow these steps, you're not going to get fire."

The four steps—or what Nester describes as a "fire ladder"—are a progression of different types of fuel that should be patiently applied as the flame and heat intensity grows:

1. Tinder (grasses, pine needles, leaves, newspaper, etc.)
2. Pencil-sized sticks
3. Sticks about the diameter of a quarter
4. Branches or logs that are forearm sized (once these have ignited then you can slowly put on larger logs)

The problem with a campfire built with only newspaper (or other tinder) and large logs is that there is not enough surface area and oxygen in the fuel materials to produce ignition. "If you are not getting fire, go back to the sequence and start over," advises Nester. "Maybe you have a lot of grass and bark in there but you don't have enough twigs. A bundle of 20 twigs has far more surface area that is surrounded by oxygen than three big logs. There is just a lot more that can ignite. Split the big logs into a bunch of smaller pieces, and then you have exponentially increased your surface area along with your chances of success."

FUEL

Bring Your Own or Collect with Care

There are two good reasons to BYOW (bring your own wood) whenever possible on camping trips. First, the immediate vicinity around most popular campgrounds are usually picked clean of dead and downed branches for fuel wood as well as forest duff for tinder. Second, it is just easier. Why not watch the sunset while relaxing next to your camp-fire instead of traipsing through the forest at dusk looking for wood? But there is a caveat: In order to avoid introducing exotic plant or insect species into a campground environment, buy wood from a local store if you are traveling from well outside the region.

"If you can bring as many of the four fire steps with you ahead of time to the campground, it is going to make life a lot easier and will reduce environmental impacts on areas that are usually just hammered," advises Nester.

The one exception is what Nester describes as "heavenly tinder" that is often there for the picking from the gravel parking lots in many campgrounds. These are the "pulverized" pine needles that have been smashed into a soft fluff by car tires. The pine resin in the needles makes the tinder extra flammable and picking up parking lot debris has no nega-

tive impact on the surrounding environment. "It is a great first step on the fire ladder," he adds.

In addition to bringing fuel wood, which can frequently be purchased at campgrounds or at grocery stores (see Chapter 4), resin-soaked sticks called Fatwood are a convenient pre-made source of kindling that serve as fire ladder step three. This all-natural fire starter can be bought in bulk boxes online or at most outdoor stores. Another easy shortcut to getting the fire going quickly, and one that can be made at home ahead of time, are newspaper logs. Simply take the black-and-white sections of the newspaper (color contains harmful chemicals when burned), roll it tight to about the diameter of your forearm and hold it in place with bailing wire. However, while the rolled newspaper will burn longer and hotter than just scrunching up pages, be aware that it will also produce large

flakes of ash that are potentially hazardous to the surrounding area.

In addition to gathering pulverized parking lot pine needles, another handy tinder trick I learned from Nester is collecting natural materials ahead of time in a paper lunch sack (see Try This, above).

In lightly used camping areas, collecting firewood (as long as it is permitted by the land management agency) is often just fine. It can even help the environment if you know how to do it properly. But, regardless of where you are, the days of stripping branches off a live tree on public lands or felling a tree for the purposes of a campfire are long gone and considered as environmentally insensitive as leaving your trash behind.

"Gather from abundance," says Nester. "Go into the tree groves that are choked with twigs and branches rather than getting wood from a location where there is just one tree or bush." However, Nester cautions against picking up fallen logs that have been on the forest floor for years and become home to plants and animals that inhabit rotting wood. Logs lying on the damp ground are also often too wet to make ideal firewood. Prime firewood comes from recently fallen branches or material that is dead and hanging above the ground.

Additionally, get a mix of hardwoods and softwoods (see All Wood is Not Created Equal on pages 44–48). Softwoods are less dense, have a lower ignition point, and are best for kindling (fire ladder steps two and three) to get the fire started. Softwood varieties that are also resinous, such as pine, are especially good for giving the flame an extra kick. But softwoods burn quickly and a fire made only of these materials will not last for long. Hardwoods are denser in terms of thermal mass and offer essential fuel that will keep your fire burning for hours.

MORE GATHERING TIPS:

- Wear gloves to protect your hands from poisonous insects, sharp branches, and splinters.
- Roam far and wide from the campsite to disperse your impact on the forest. And gather all the wood you will need well before dusk to avoid walking around in the dark.
- A general rule is that an armload full of wood will keep a fire going for about 45 minutes. Either bring enough fuel from home or gather from the forest during the day to meet your desired evening campfire activities.
- Strive for dry, seasoned wood. If it is brittle enough to make a snapping noise when you break it, the twig or branch is a keeper.
- Avoid green wood, which does not easily snap when you bend it. A live tree is more than half water, so wood that was recently growing is very wet and hard to burn. It also produces a lot of smoke.
- Long-dead wood that is crumbly is also too damp to be of use. If you can easily crush the log's surface with your finger it is too rotten to burn.
- When gathering fuel logs, smaller diameter branches (no bigger than calf-sized) heat up more efficiently than larger pieces, which often don't burn completely and leave a mess in the fire ring.
- Rather than attempting to break apart downed branches with your hands, bring a small handsaw along to do the work for you and avoid injury.

All Wood is Not Created Equal

Your geographic location will determine what types of natural tinder, kindling, and fuel wood are available to feed your campfire. Here are some of the best options for campers in the United States and Canada.

TINDER

This first step on the fire ladder needs to be dry and fluffy. Before lighting it you should craft it into a palm-sized bundle that is slightly cupped like a birds nest. You can place your match in the cup. Some natural tinder materials that are generally available throughout North America are:

- Dead grasses and leaves clumped together loosely to allow oxygen to circulate
- Conifer needles and cones from pine, spruce, cedar, cypress, larch, and hemlock trees that are on the ground. Grab needles still attached to the node because this tip is infused with flammable resin.
- Loose bark from cedar, mulberry, juniper, aspen, cottonwood, poplar, and oak. Rub together between your hands to make it soft.
- Cattail fluff from marshy and riparian areas that is most abundant in the summer and spring. A handful lights in a flash and disappears just as quickly.
- Dandelion seed heads are as flammable as cattail fluff. Fill a few bags of fluff for your next

camping trip to make use of these pesky intruders in your lawn that are most prevalent in the spring and early fall.

There are more regional tinder materials too. In the northern U.S. and Canada, birch bark is highly flammable and easy to retrieve from dead, fallen branches. Also, be on the lookout for something known as tinder fungus that grows on live birch trees in black bulbous blotches. The inside of this fungus is reddish brown and quickly catches a spark, continuing to burn longer than grasses and pine needles. In the southeastern U.S., Spanish moss that hangs from trees, sometimes called an "air plant," is dry and fluffy and loves to burn. Along the Gulf Coast and in the deep South, palm tree bark at the base of palm branches (fronds) makes outstanding tinder. In the southwestern U.S. and Great Plains, snag one of the tumbleweeds blowing down the road. The bark from these dead sagebrush bushes can be pounded with a rock to produce a pile of shredded wood that is highly flammable and also smells great.

TRICK FIRE STARTERS

Are you out of pine needles but you have a bag of Fritos in the car? You're in luck! Here are some of the favorite trick tinder sources used by outdoor survival experts, listed in order of effectiveness:

Vaseline cotton ball: Thoroughly coat a cotton ball in Vaseline, and then set it ablaze. Editors at *Backpacker* magazine report this burns for a full eight minutes.

Steel wool and a 9-volt battery: Fluff up a piece of fine steel wool and touch one end to the positive node on the battery and the other end to the negative node. This will produce a spark and the steel wool will burn for several minutes.

Dryer lint: Loosely balled-up lint is readily flammable, but for extra punch drizzle the lint with candle wax before lighting.

Fritos: Thanks to the healthy coating of vegetable oil, a Frito will burn for about a minute. Lays potato chips and Doritos will also work but not quite as well Fritos, which are very dry and high in saturated fat.

Magnifying lens: The bigger the magnifier the better, but even powerful reading glasses can do the trick. The goal is to concentrate enough heat from sunlight to get tinder to smolder and then combust. Hold the glass very still over dry tinder and adjust the height and angle to get maximum benefit from the sun.

Spent lighter: When a disposable lighter is out of fuel, it is still a potentially useful tool for fire starting. Try cramming dry tinder into the empty fuel cylinder but don't let it get in the way of the wheel/flicking mechanism. This works best if the tinder is soaked in something highly flammable such as camp stove fuel or even pine resin.

KINDLING

Whether you are bringing it from home or gathering it around the campsite, the best kindling comes from softwoods. In the United States, these are coniferous, evergreen species such as pine, spruce, and fir that are infused with resin making the wood quick to burn. But it is also quick to burn out.

"All the outdoor experts I have ever worked with get their fire going with softwoods because of the low ignition point," says Nester. However, be prepared to switch to hardwoods (see page 47) once the kindling is burning in order to establish a longer-lasting fire.

Gathering plenty of twigs and sticks from the ground will not only help get your fire started, but it is also the best way to increase heat later if the fire starts to fizzle and fuel logs are not igniting. The second step on the fire ladder should be twigs that are about the diameter of a pencil. Once these are burning you can add the third step, which is comprised of larger sticks about as big around as a quarter. An easy way to carry kindling and place it in the fire ring is to gather the twigs in your hand like a bouquet of flowers. Hold the thicker ends of the sticks at the bottom with the wider, forked ends at the top. Use a piece of twine to hold the bouquet together. Then stuff the top of the twig bundle with fluffy tinder. Place the bouquet upside down in the middle of the firelay structure (see pages 50–52) and apply a match to the base.

FUEL WOOD

If you have faithfully followed fire ladder steps one, two, and three, and a spark has lit the tinder, which

---

WHAT ABOUT WHISKEY?

Backpacker magazine editor Casey Lyons reports that besides the pleasure of drinking it on camping trips, there are two important ways whiskey can come in handy for fire building. First: Whiskey that is 80 proof or stronger is a powerful flame accelerant like lighter fluid. Pour a generous amount on your tinder and light it to get instant combustion. Second: If your whiskey bottle is empty you can use the bottle itself as a fire starter. Fill the bottom of the bottle with water then angle it over the tinder so that it acts as a magnifying glass. Hold the bottle steady (assuming you have not drunk too much whiskey) until the hottest part of the sun's rays directly overhead produces smoke and then flame on the tinder.

If the sticks on the ground are wet or you just want to enhance the flammability of your kindling, shave the exterior of a few thumb-thick pieces of wood with a pocketknife to create what is called a feather stick. Being careful to stroke the knife blade downward and away from the body, hold the stick at its base and shave away wet exterior bark to expose the dry inside on the top half of the stick. Shave down layers that are thin enough to curl. If the stick is wet, pull off the dried shavings for the fire and discard the rest. Or place the entire dry stick into the fire on top of the tinder. Compared to a regular stick, a feather stick helps a fire ignite because of this kindling's increased flammable surface area.

then ignited a pile of kindling that significantly increased the heat of the fire, it is time to add the fuel wood.

This fourth and final step of successful fire building requires hardwood. Unlike low density, resinous softwoods, hardwoods are, as the name implies, harder and heavier with tightly packed fibers. Hardwoods are much tougher to light but they provide long-burning, consistent heat. In North America, hardwoods are generally deciduous trees such as hickory, oak, walnut, maple, birch, aspen, and cottonwood. However, there is one notable exception: juniper. This evergreen coniferous tree generally found throughout the West burns long and hot, and makes excellent fuel. (See Chapter 6 for the BTU burn rates and other properties of different types of hardwoods.) While it may be tempting to haul giant fuel logs into camp, stick with logs that are the diameter of your calf or smaller for the average campfire. These are much easier to burn completely for maximum heat output. If you have an ax, you can also split larger logs into more manageable and flammable pieces (see log-splitting instructions on pages 129–30).

In camp assemble the four categories of fuel into

separate piles. Make sure your tinder and kindling is protected from the damp ground by placing it on a tarp or plastic trash bag. Keep the materials several feet away from the fire ring but where you can easily reach exactly what you need when you are sitting around the flames.

FIRE

Now that you have all the right stuff, it is time to assemble the materials into a design that effectively brings together the three essential components of the fire triangle: heat, fuel, and oxygen (see Chapter 1).

But first a note about the location of your fire: A suitable site should be in a place that is relatively sheltered from wind but not surrounded by dense vegetation. An opening in the forest canopy is preferable but if there are tree branches overhead they should be at least 15 feet high. Campfires have significant long-term environmental impacts because the heat sterilizes soil underneath and blackens rocks. Consequently, it is best for the forest to build your fire in an existing fire ring where the damage is already done. However, if you don't find the tell-

tale signs of a circle of rocks and partially burned logs at your campsite, try to build a new fire ring in a location that has mineral soil (i.e., sand, gravel, or rock) rather than rich, black dirt. (See pages 57–59 for more ways to reduce environmental impact by building a pan fire or mound fire.) Before you assemble the fuel, clear the fire pit and surrounding area of flammable debris. Then dig out a slight depression in the pit that is about as big around as a beach ball and two inches deep. This depression allows the coal base to concentrate at the bottom to keep the fire going, but it is not so deep that it prevents airflow.

Built to Burn

You could just haphazardly throw everything into the fire ring and light a match, but the odds are not good for getting ignition once the tinder burns out. That's because just as starting a car requires a sequence of mechanical steps, so does a campfire. The way the materials are placed in the ring must send the flame from the tinder to the kindling and then to the fuel, building heat and momentum with each step.

Here are some of the most tried and true fire designs, often called firelays because they describe how the fuel should be placed on top or around a pile of lit tinder. Make sure you have a jug of water close at hand in case flames jump out of the fire ring.

TEEPEE

By far the most commonly used firelay, the teepee is popular for the way it produces flames quickly with intense heat rising from the center to warm cold hands. "The teepee design is great because of its strong updraft, which reduces smoke and encourages combustion," notes Nester. The downside of this design is that the teepee can collapse before the fuel wood is fully ignited causing you to have to start the fire-lighting process over from the beginning. Try to prevent early collapse by creating an exoskeleton as the foundation of the teepee. Use a kindling stick that is about as long as your forearm and as thick around as your thumb with a fork on one end. Drive the other end of the stick into the ground at about a 45-degree angle over your tinder bundle. Then position a similarly sized stick in the crook of the fork to form the opposite side of the teepee. Next, arrange smaller sticks (ladder step 2) around the structure but leave an opening where you can access the tinder bundle in the center to light it with a match. Once the tinder is lit, begin gradually adding larger pieces of kindling (ladder step 3) to the walls of the teepee. Keep feeding the fire with kindling until you have some hearty coals at the base. Then you can place smaller pieces of fuel wood around or over the coals, but don't cut off airflow to the coals or your fire will die out.

LOG CABIN

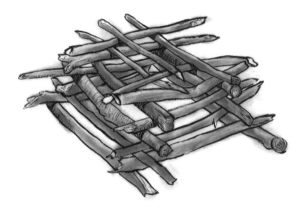

Although harder to get going initially, this square firelay (envision a house made of Lincoln Logs) is a great choice for large social gatherings around a campfire where long-burning heat and light are desired. Place two smaller pieces of fuel wood (several inches in diameter) running parallel on either side of your tinder pile. Then place two sticks of kindling (ladder step 3) across the fuel wood to form a square base for the log cabin. Next, place a few kindling sticks diagonally across the square but leaving enough space to insert a match to the tinder below. Continue to build the walls of the square structure with alternating smaller and larger kindling until it is about 7–9 inches tall. Light the tinder and continue to add more tinder and kindling to the center of the structure to feed the flame. As the fire heats up slowly add fuel wood to the square firelay. This design is also an easy way to produce coals for cooking or s'mores. Once the fire is going strong push some coals into a corner of the square and then nudge them just outside the "cabin" where you can safely access them for marshmallow roasting. (See Chapter 7 for more fire cooking tips.)

PYRAMID

If it's windy and/or drizzling, this firelay offers the unique benefit of sheltering the flame while also feeding it. Similar to the log cabin, assemble the base in a square around your tinder pile. Start with a foundation of small fuel logs (wrist-sized), then add rows of sticks atop the tinder, with the sticks spaced about a half-inch apart. Each row should run perpendicular to the preceding one and use progressively smaller sticks—hence the pyramid design. Make sure you can access the tinder with a match, and once you light it, slowly add more sticks to the pyramid. If conditions are blustery, use kindling from resinous trees such as pine and spruce to boost the burn factor. After there are coals at the base, gradually overlay the pyramid with fuel logs.

LEAN-TO

In windy situations, such as at the beach, this design is often effective. Push a fuel log that is about wrist thick into the fire pit at a 45-degree angle. Position the log so that the upper end is pushing into the direction of the wind. You may also rest the log against a rock on the fire ring. This way the log is both a fuel support structure and a windbreak. Place the tinder pile under the log and then lean alternating small and large kindling pieces (ladder steps 2 and 3) about one half inch apart on either side of the log. Light the tinder and slowly feed the flame with more kindling. Continue to use larger pieces of fuel and the walls of the fire ring as a windbreak.

STAR

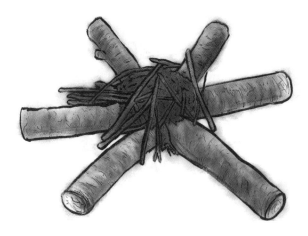

While it is ideal to have a wide variety of sizes of fuel wood and kindling, sometimes you may find yourself stuck with a bunch of tinder and a few big logs. In this situation, you can assemble the fuel like spokes on a wagon wheel where they intersect in the center on top of a tinder pile. It takes patience and a lot of tinder to get this started but once the fuel logs are burning at the hub where they intersect, continue to move the logs in toward the center to keep the fire going.

Care and Feeding

TURNING UP THE HEAT

Contrary to popular belief, throwing a big log on a fizzling fire is not the best way to boost BTUs. The quickest method for turning up the thermostat is to throw small and large kindling pieces directly on top of the blaze. The sticks immediately ignite and draw convective heat from the bottom of the fire where the coals are simmering and up toward the sky. The rising heat makes the top of the fire the hottest part, but only while the kindling is burning. If you want to keep the fire warm and cozy for hours, periodically add fuel logs to the base along the outside edges where there are coals but not tall flames. Feeding the cooler parts of the fire with hardwood makes for a longer lasting blaze that gives off widely dispersed heat. However, don't make the fire bigger than you need. The more wood that you burn, the more water and time will be required the next day to properly extinguish the coals. (See pages 59–60 for how to put out a fire). For fires that are annoyingly smoky, try moving around fuel logs to create a stronger updraft. If you are still getting a lot of smoke, your wood is likely wet or green.

WHEN IT'S WET

It's definitely possible to start a fire and keep it burning in the rain. First, you must make sure your tinder and kindling are dry. If you gather wet materials, carry the tinder and some kindling under your jacket to allow your body heat to dry it

STAYING WARM

When people get stranded in the woods, a common fear is that they can't survive without food. But the fact is, while you may be very hungry, it actually takes weeks to starve to death. A far more common cause of death among people who get lost in the wilderness is hypothermia. Even in summer the body can become dangerously chilled at night, especially if conditions are damp. In winter, or anytime when temperatures are below freezing, people can sometimes die within a few hours if they don't have a way to keep themselves warm. In this regard, fire can be a true lifesaver.

No one plans to get lost, but you can plan to be properly prepared and equipped if you do find yourself hopelessly turned around. Always carry three different means of starting an emergency fire on outdoor excursions. This should include a lighter, some waterproof matches, and a low-tech method such as a magnesium striker or magnifying glass. Another handy stormproof fire starter is a 9-volt battery and steel wool. Fluff up a piece of steel wool and place two edges of the wool to the two battery contacts. This will create a circuit and the steel wool will start to spark burn. And because steel wool is excellent tinder, it will burn for several minutes while you use it to get your fire going.

When building a fire to survive the night, keep it small and positioned near some type of overhang or object such as a cliff face or large hollowed out tree. The goal is to not only warm yourself with the fire but also with the reflective heat coming off the cliff or other object. While emergency space blankets are fairly useless for providing warmth, they are excellent for tying to tree branches to create a reflective wall. In addition to three forms of fire starters, you should always carry in your car and in your pack a space blanket for this purpose as well as some of the "trick" tinder detailed below. Build the small fire about 10 feet away from the reflective wall. Then gather pine needles or other duff and shape into a bed between the wall and the fire. You want to make a place for yourself that is off the damp ground and able to catch the heat coming from both directions.

SIGNALING FOR RESCUE

Whether it is day or night, fire can be an effective means of alerting aerial search and rescue crews to your location. However, unlike the fire built in a sheltered area for staying warm, a signal fire should be in the most wide-open place possible. A large meadow, hilltop, or ridge is ideal. The universal distress signal is anything in a pattern of three, so a clear SOS would be three fires each about 20–30 feet apart forming a triangle. During the day, you want to draw attention with smoke. Grasses and wet wood can be extra smoky or wood with resin that creates black smoke is more noticeable against the sky than white smoke. Be judicious in deciding whether or not to start signal fires. If it is windy the smoke will disappear and not be noticeable, and you don't want to compound your problems by causing a wildfire.

out. Kindling can also be dried out by placing the sticks near—but not in—the fire. Also, carve a few feather sticks (see page 47). If you have wet fuel wood, split it in half with an ax and place the dry, inner sections on the flame. Make your firelay a log cabin or pyramid design (see pages 51–52) and use resinous sticks for kindling. Once the fire is well established, stack wet or green logs up against the exterior of the square structure (allowing a few inches of space for air circulation) and let the heat dry out the wood before you place it on the coals. Use a lot of smaller pieces of fuel wood instead of larger logs to increase the surface area of flammable materials and build up a thermal mass that is impervious to dampness.

FIRE AND ICE

In situations where you are winter camping and there is snow on the ground, fire is still possible, although you'll likely need to bring your wood with you. Make a platform for the fire by tramping down the snow into a firm floor. Then build a raft on top of the snow with green wood that will provide a dry base for your fire ring. Assemble a layer of logs that are approximately three feet long and placed side by side. Then make a second layer running perpendicular to the first. Jam a fuel log about the width of your forearm into the snow and up against the raft to act as a brace against which you can place your kindling. Using the lean-to firelay (see page 52), the brace allows oxygen to circulate between the kindling and the raft base in order to keep the fire burning. Because the base is green, moist wood and it is on top of snow, it is unlikely the platform underneath the fire will burn. Another simpler and more environmentally friendly method for enjoying a fire in the snow is to build it in a pan (see page 57).

ENVIRONMENTAL CONCERNS

While fires are one of the most enjoyable parts of a camping trip for many Americans as well as a time-honored outdoor tradition, having one on public lands is becoming increasingly problematic. Many land management agencies restrict or ban campfires in popular places in order to protect against the depletion of natural resources. And in places where campfires are normally allowed but recent drought has made the landscape dangerously flammable, fires are also often banned during certain times of year. Be sure to contact the land management office overseeing the camping area you will be visiting well in advance of your trip to find out about campfire restrictions.

If campfires are allowed but you want to be sensitive to environmental impacts, follow the guidelines of Leave No Trace. First developed by the U.S. Forest Service in the early 1990s, Leave No Trace principles are a kind of outdoor etiquette that encompass ways to minimize environmental impacts of campfires as well as trampling on trails and campsites, littering, disturbing wildlife, and infringing on the outdoor experience of others. The LNT program is now embraced by all land management agencies in the United States and practiced by the Boy Scouts of America, the National Outdoor Leadership School, Outward Bound, and most other outdoor groups. Like all LNT principles, which encourage hikers and campers to "take only pictures, and leave only footprints," the program does not prohibit campfire use but, instead, teaches ways to have a fire without creating lasting impacts. One approach is to only have fires in pre-existing fire rings where historic use is well established. But what if there is no fire ring?

Campfires the LNT Way

When camping in a pristine area with no sign of previous campfire use, the best method for building a minimum-impact fire is to do it in a self-contained unit such as a fire pan or on a mound. When done properly, neither the pan fire nor mound fire should have any impact on the soil beneath it. And, once the fire is cleaned up, there will be no evidence that it ever happened.

PAN FIRE
A technique perfected and widely used by river runners, the pan fire offers the easiest and most envi-ronmentally sensitive way to get that warm glow in the backcountry. Fire pans are simply anything that serves as a flame-resistant metal tray with sides high enough (at least three inches) to safely contain wood and ashes. Various common household items work fine—oil drain pans, garbage can lids, and pans from backyard barbecue grills. Now that fire pans are becoming increasingly popular among backpackers, several outdoor equipment companies have begun to market highly packable, lightweight fire pans. There is not much to making a pan fire—just fill the bottom of the pan with a few inches of sand and assemble the fuel firelay. Prop the pan up on a few rocks to protect the surface below from any heat damage.

MOUND FIRE
This type of fire has its advantages in that you do not have to bother with packing a bulky fire pan; however, it also requires more work in camp. The tools needed are a trowel, a large stuff sack, and a ground cloth.

First, locate a source of mineral soil for building the mound; never build it on dark organic soil because this kind of dirt is rich with living organisms. Some of the best places to find mineral soil are dry streambeds where gravel or sand is easily accessible and the soil is frequently disturbed by flooding, or beneath the turned-over stump of a fallen tree. Using the trowel, fill the large stuff sack (you can turn it inside out to keep it clean for other

uses) with the mineral soil. Make a special note of your dig site so you can replace the soil when you tear down the fire. Then return to your campsite and lay down a small tarp or ground cloth (a plastic trash bag also works fine) on the spot where you intend to have the fire. Spread the soil on the cloth, forming a circular, flat-topped mound about 6 to 8 inches thick. The thickness of the mound is critical in preventing

any heat-caused damage to the surface beneath the ground cloth, and to keep the ground cloth itself from melting. The circumference of the mound should be larger than the fire to allow for the inevitable spread of coals.

Both pan and mound fires should be kept small, using only small scraps of wood for fuel. And the wood should be burned all the way down to a fine ash to eliminate the lingering presence of charcoal. When you are ready to pack up camp, make sure the fire is completely out and then scatter the leftover ashes across a broad area away from the campsite (the ashes should be cool enough for you to run your hand through). If you built a mound fire, return the mineral soil to its original location. For more on Leave No Trace practices, go to: www.lnt.org.

Keep Smokey Bear Happy

On June 20, 2010, the embers from an abandoned campfire in northern Arizona's Coconino National Forest took flight and ignited surrounding brush. When the first fire crew arrived at the location near Schultz Tank north of Flagstaff, the blaze encompassed two acres. But 50 mile-per-hour wind gusts soon overpowered the crew. Over a period of ten days, the fire continued to grow and consumed more than fifteen thousand acres of national forest in the heart of the San Francisco Peaks, one of Arizona's most popular recreation areas. One month after the Schultz Fire was finally put out, torrential rainstorms in the burned area caused flash floods because the fire had stripped away topsoil that normally acted like a sponge. Some thirty million gallons of rainwater, along with a torrent of ash, debris, and boulders as big as Volkswagens rolled off the Peaks and into residential areas below. The flash floods engulfed homes that had been evacuated during the fire just weeks before and a 12-year-old girl walking on a neighborhood street was swept away and drowned. All this tragedy could have been prevented if the campers had properly extinguished their campfire.

A recent study of wildfire data from 1992–2012 found the vast majority of wildfires in the United States are caused by human carelessness. Beyond the $2 billion per year cost of fighting them, the fires resulted in millions of dollars in property damage as well as the loss of life.

The U.S. Forest Service first introduced its Smokey Bear Wildfire Prevention Campaign in 1944 to educate campers on how to properly extinguish a campfire. The goal was to get increasing numbers of recreationists to take personal responsibility for preventing wildfires and to understand what it meant to put the fire "dead out." Even after all these years, the Schultz Fire in Arizona and many other wildfires started by campfires prove that Smokey Bear's message is as relevant as ever. Even when there is no longer wood burning in a fire ring or visible signs of smoke or flame the morning after a campfire, a deep bed of coals is often still smoldering beneath a layer of ash. Just a little bit of wind can stoke the coals and carry sparks into surrounding vegetation. A fire that is "dead out" means the coals are completely extinguished. Depending on the size of your fire, accomplishing this requires a fair amount of water and elbow grease.

Before starting the extinguishing process, let the fire burn all the way down to ash. Pour several

liters of water onto the ash pile and then go about your other camp chores for an hour or two while the coals cool. Next, finish up the job with two people: One person should pour water into the fire ring while the other person uses a shovel or a walking stick to poke and stir the coals. It is critical for water to completely penetrate the coal base. Keep pouring and stirring until the ash and what is left of the coals are cool enough to touch with your hand.

This process can be a water-intensive undertaking, so make sure you bring plenty to complete the job. Remove unburned pieces of wood and coals and pack them out with your trash. Take the cool ash in the fire ring and disperse it widely around the perimeter of the campsite. This way, the fire ring will be an inviting place for the campers who come after you and reduce the chances of a new fire site being built.

FIRE KEEPER: TONY NESTER

Like many people who make their living in a career that keeps them outdoors, Tony Nester first got his introduction to sleeping under the stars from the Boy Scouts. "I had some great Scout masters who were crusty old woodsmen," says Nester. "They took me fishing and camping and taught me all kinds of outdoor skills."

Nester's fascination with the outdoors grew during his high school years as he read the nature writing of Henry David Thoreau and Ralph Waldo Emerson. The idea of being self-reliant and living off the land appealed to Nester, so after high school he decided to give it a try. He embarked on a solo wilderness trek in Michigan's Upper Peninsula with almost no gear and little food, intending to live simply as Thoreau romanticized in his book *Walden*. "The rule is to carry as little as possible," wrote Thoreau. And that is what Nester did.

"I jumped off the deep end. I was totally miserable and cold the whole time," says Nester. "I had such a negative experience on that trip that I gave up the outdoors and worked at a sporting goods store instead."

However, after three months of never leaving the city, the wilderness called to Nester again. Only this time he decided to take all the backpacking gear he could possibly carry in order to be warm and comfortable. With a 60-pound pack on his back, Nester did a solo expedition in Michigan's Porcupine Wilderness where he relished living in isolation and doing things like following bear tracks through a swamp. And while he was glad to not be cold, he also found all the gear to be a hindrance to his wilderness experience. "I felt detached from the environment when I was carrying so much stuff," he explains. "It was a barrier between me and the natural world. So when I got back I started looking into survival schools."

Nester took all the courses he could find, not only in wilderness survival techniques, but also in edible plants and primitive living skills (also called bush craft). Through this increasing knowledge he was able to find a way to be comfortable and connected to the outdoors without carrying large amounts of gear. In 1989, Nester founded his own outdoor survival and primitive technology school called Ancient Pathways in Flagstaff, Arizona. Over nearly three decades, he has trained members of the U.S. military, National Park Service rangers, and even Hollywood movie stars who needed to act in authentic wilderness survival scenes.

One of the primary values Nester tries to instill in students is an appreciation for fire. "When I teach students how to start a fire with a hand drill I emphasize that we are here today because our ancestors perfected this technology," says Nester. "Maybe you don't really need to know how to use a hand drill, but fire is our common denominator. Whether you are Native American or Swedish, we all had ancestors who rubbed two sticks together. Fire is not just an aboriginal skill; it's a human skill."

Campfire Alternatives

If you find yourself pitching a tent in a place where campfires are not allowed or you just don't feel like building one and cleaning up afterward there are plenty of illuminating alternatives. Here are a few ideas for faux fires:

CANDLE LANTERNS

When enclosed in a lantern that amplifies light, the flame of a single candle can provide enough light for card games or just fending off the darkness. An ingenious outdoor gear company based in Washington called UCO (Utility, Comfort, Originality), offers a wide array of excellent candle lantern options.

LED LIGHTS

Make your campsite festive any month of the year by stringing up portable LED lights that can be attached to trees or tied above hammocks. One option is Twilight Camp Lights from Eagles Nest Outfitters, which contain 23 tiny lights that are powered by three AAA batteries for 72 hours of continuous burn time.

LIGHT UP A COLORED WATER BOTTLE

Shining a headlamp into a nearly full water bottle produces a comforting glow that disperses light and color several feet in every direction. Fill a water bottle 3/4 full, then strap a headlamp over the open top with the bulb facing down. Loosely place the bottle lid on top of the lamp to send the light downward.

VIRTUAL FIREPLACE APP

Why not? Gather around in your puffy jackets and sleeping bags, fire up the camp stove to make hot chocolate, and place your glowing tablet in the middle of the circle. A variety of apps from Google and Apple provide nonstop visuals of burning logs along with sounds of crackling wood. The only thing missing is smoke in your face.

CHAPTER 4

Backyard Fire

A well-positioned fire pit can turn a backyard into a fun gathering spot for family and friends

Is there anything that can compete with video games when it comes to captivating the interest of adolescent boys? During my son's teenage years, it seemed that he and his friends spent much of their free time gathered around a TV screen and game console with controllers in hand. Regardless of whose house they were hanging out at, the activity was always the same.

But then the father of one of my son's friends built a fire pit in the family's backyard. And much to the surprise of myself and the other parents, our boys often opted to sit around the fire on lazy summer nights instead of playing video games. Once away from the flickering TV screen, they gazed into real flames with their feet propped on rocks, throwing big logs on the fire as they told jokes and pondered the mysteries of middle school.

In ancient times, language, cooking, and togetherness evolved while humans sat around a shared fire at night. Now the glowing television has become something of a pseudo family campfire in our culture as it is the central attraction of the modern living room. But it is a poor substitute for the real thing. Just as my son and his friends discovered, a fire is captivating in a way that TV is not; it fosters conversation, shared experience, and a general sense of contentment. However, many people have relegated outdoor fires to something that only happens when camping. And in our normal, day-to-day lives most evenings are spent indoors—too often watching TV. The backyard may have a barbecue pit and some lawn furniture but it is not considered an evening living space.

However, things are changing. One of the most popular trends in landscape architecture today is the creation of the "outdoor room." These comfortable living spaces in the backyard don't have walls and a ceiling like traditional rooms but they almost always have a fire feature as the centerpiece in much the same way the entertainment center is at the heart of the modern family room. The flames can come in the form of the traditional word burning fire pit but may also be a natural gas or propane-fueled fire pit. Some outdoor rooms have fireplaces, chimeneas, or hearths. And for ambitious cooks, there may even be a wood-fired oven.

Deciding to bring the pleasures of an evening fire into your backyard can be as simple as buying a portable fire pit or as involved as designing an outdoor room complete with a new patio. Either way, there are some basic regulatory and safety factors you should first consider.

Playing It Safe

REGIONAL FIRE RESTRICTIONS

Rules on open burning—which is any combustion outdoors that is not filtered through a stovepipe or chimney—vary widely between geographic regions in the United States and also by city. Arid regions in the western United States have strict regulations that sometimes ban burn wood burning fires year-round or during the driest times of the year. And when outdoor fires are allowed in the West, municipal regulations often require that the fire pit be covered with a protective screen. Other locations in the United States that receive more moisture have fewer restrictions on backyard fires. In areas where wood-burning outdoor fires are banned for reasons of air pollution and/or the threat of wildfire, a natural gas or propane-fueled fire pit is an excellent option (see alternatives on pages 71–73). Before purchasing backyard fire features, research the regulations in your area. Fire restrictions are usually clearly posted on municipal websites.

LOCATION OF FIRE SITE

While you want your "outdoor room" conveniently located in relation to the rest of the house, a backyard fire pit should be at least 10 feet away from any structures. The fire should also not be under overhanging tree branches, even if they are 50 feet high.

"The convective heat column rising from the fire needs a clear path so the heat can dissipate and cool," explains Mark Shiery, a retired municipal and wildland fire fighter. "The branches create a lid that captures heat, so it is much safer to have open sky above a fire." Shiery notes that embers escaping from the fire pit and flying through the air are a hazard, but the heat itself that can transfer to vegetation

A chimenea is a free-standing outdoor fireplace.

above also kills branches, which are then more prone to catching fire.

FORTIFY THE AREA

"The same situations that cause wildfires in the forest can happen in a backyard," says Shiery. "You need to get rid of receptive fuels where embers can land and start spot fires. And the drier the air, the longer the embers stay lit and travel." When he was fighting fires on western national forests, Shiery witnessed embers traveling more than a mile and starting new fires. He says embers can land in backyard gutters full of pine needles, flowerbeds full of wood chips, or accumulations of leaves under a deck. Green grass is generally safe, but the backyard should be thoroughly raked and all dead debris removed before having a fire. In addition, the surface underneath and around the fire pit should be made of a naturally fireproof material such as gravel, concrete, stone, or brick. Having a patio that is dedicated to the fire pit not only serves as a durable surface for groups to gather but also provides protection against the fire getting out of control. Shiery adds that furniture used around the fire pit—including chair cushions—should be of the non-flammable variety.

Wood Burning Fire Pits

For all the same reasons that a fire is one of the most beloved parts of a camping trip and a crackling fire in a fireplace adds ambiance to an indoor

living room, a real wood burning fire pit in the backyard is hard to beat (as long as it is legal). Compared to natural gas or propane, a wood burning fire has the ability to give off a more intense heat. Firewood is readily available and offers aesthetically pleasing aromas. Plus you can cook on the fire and the coals. And as my son and his friends discovered, there is something pleasingly primal about sitting around dancing flames generated by a technology that has been used by humans for more than a million years.

Portable wood burning fire pits are basically large metal bowls that range in diameter from about 30-36 inches and are about 20-28 inches deep. Most have legs to keep the container a few inches above the ground and all have tight fitting spark arrestor screens to prevent flying embers. The bowls are made from stainless steel (which is the least expensive), cast iron, or copper. Cast iron is durable and has the added benefit of radiating heat from the bowl's surface. Deeper iron bowls often have cut-out designs on the walls that add interest and allow viewing the fire from the side as well as the top. However, iron is prone to rusting and needs to be covered when not in use to protect against moisture. Copper is visually appealing and can add an aesthetic element to backyard landscaping but it is also significantly more expensive. Copper does not rust and it gains a pleasing patina over time. No matter what they are made of, portable fire pits should be placed on concrete or stone and never on wood decks.

The main advantage to a portable fire pit, besides the fact that it is portable, is the tight-fitting spark-arresting screen and the ease of cleaning. The bowl can just be picked up and the cold ashes dumped into the garden or trash. But these pits are generally smaller than permanent fire features and they are not integrated into the backyard space the way a permanent fire pit is.

Building a fire pit in the backyard can be as simple as following these steps:

Step 1: Find a level patch of ground that is at least 12 feet in diameter. Drive a stake in the ground that is attached to a string 2–3 feet long. Rotate the string to form a perfect circle and mark the perimeter with spray paint. Remove dirt from the circle, reaching a depth of about 18–20 inches deep.

Step 2: Shovel 6 inches of fine gravel into the pit and rake it smooth. Tamp down the gravel to create a solid base. Then spread a layer of coarse sand that is 1–2 inches thick. Tamp it down and make sure the surface is level.

Step 3: Cut a strip of sheet metal to match the depth and circumference of the pit. Wrap the metal around the inside of the pit, covering exposed dirt. Use iron stakes to secure the metal flush against dirt wall.

Step 4: Place concrete retaining wall blocks around the rim of the pit. Use a rubber mallet to pound blocks into place and make sure they are flush with one another. Lay a second layer of blocks that is staggered over the first layer. Attach the two

tiers of blocks using masonry adhesive. Allow slight gaps in the blocks to encourage airflow to the fire. Allow two days for the adhesive to dry before you use the pit.

For building a fire pit along with a patio, see "making a place for your pit" on pages 74–75.

Alternative Fuel Fires

NATURAL GAS AND PROPANE

Unlike wood-burning fire pits, backyard fire features that are fueled by natural gas or propane are widely allowed across the United States because they do not contribute to particulate air pollution or emit embers that can start wildfires. However, some extremely dry cities in the West that are especially vulnerable to wildfire require that natural gas or propane burners be registered. Rather than have fake wood, most of these pits are filled with lava rock that covers a circular gas burner. Some fire pits are filled with colorful glass or specially heat-treated river rocks. It is important to only use fire pit materials that have been provided by the manufacturer and are designed to withstand extreme heat. Untreated river rocks or stone may explode and cause serious injury to those sitting around the fire pit.

If you are contemplating upgrading the landscaping in your backyard, a stationary natural gas fire pit can be incorporated into the design project. The pit will need to hook into a natural gas line and be located in a place that follows local building codes. When constructing a patio with a gas fire pit, it is also convenient to use the same line for hooking up a gas barbecue grill as part of the outdoor room. Before any construction begins be sure to have all utility lines marked on your property. Most regions in the United States are part of the North American One Call Referral System (888-258-0808), which will contact all the utilities in your area to have the

TRY THIS: TURN AN OLD WASHING MACHINE DRUM INTO A FIRE PIT

If you're feeling handy or just like the ethics of up-cycling, head to a junkyard to find your fire pit. Your quarry is the drum from any top loading washing machine. The size, shape, and sturdy stainless steel construction with its many tiny holes make the inner drum a perfect portable fire pit.

- Extract the drum from the body of the machine.
- Remove the plastic rim around the drum and all other non-metal pieces.
- Use an angle grinder to cut out the drum's center spindle.

- Use the grinder to cut off the metal lip on the top of the drum and then smooth the surface and polish away any soap scum. (This is for cosmetic, not functional, reasons.)
- You can choose to make legs for the drum with metal pipe or by cutting legs off an old metal bed frame and welding them to the bottom. Or you can go without legs and place the drum on a foundation of firebricks (regular clay bricks or other stone could dangerously explode under the fire's heat).
- Paint your drum a fun color with high temperature paint, and you are done.

lines in your yard marked. The network generally requests at least three days' notice before you plan to begin excavation.

Propane fire pits are usually mobile and hook up to a propane tank that is not far from the pit. Often the pit is incorporated into a table that offers a place to prop feet and also hides the tank. Compared to natural gas, the mobility of the propane-fired pit offers more flexibility for backyard uses and can also be an easy back-up for times of year when wood burning fires are banned. The drawbacks are that the fire is generally smaller than a natural gas pit and the tank can run out at inopportune times.

Even though natural gas and propane fire pits do not generate embers, they still emit steady heat that travels up in a column. In order to provide proper venting, they should be situated under the open sky without tree branches overhead and should also be at least 10 feet away from any structures. A valve and/or push-button ignition makes these fire pits easy to start and instantaneous flames can be adjusted by regulating the gas flow. There is also no mess of ashes to clean up when the party is over.

ECO-FRIENDLY FUELS

Bioethanol is the latest innovation in clean burning fire technology. Portable fire pits using bioethanol are completely free standing and are allowed throughout the United States just like natural gas and propane units, but these come with a smaller carbon footprint than fossil fuels. Bioethanol is a clear, odorless liquid produced from fermented plant material, such as sugar cane, corn, and wheat. Currently, most bioethanol fire pits utilizing this new and evolving technology have fake ceramic logs that are fed by a liquid fuel canister attached to the fire bowl. A one-liter canister burns for approximately two hours.

Logs made from alternative materials such as coffee grounds, grass, and soy also offer environmentally sustainable options for a backyard fire pit. Yet, even though these innovative "logs" are not made from wood, they are still restricted in places where wood burning fires are banned. The Pine Mountain Java-log is made from recycled coffee grounds and burns for four hours. Unfortunately, it does not smell like coffee when it burns but it does give off BTUs comparable to wood. Pine Mountain, the company that produces the Java-log, has established agreements with large coffee shop chains to pick up their discarded coffee grounds and turn the garbage into fuel. Coffee grounds contain more oil than wood, which makes it easy to burn, even though it has less carbon than trees. According to OMNI, an independent testing company, the Java-log burns seven times cleaner than wood and emits 96 percent less residue into the air. It comes in a paper wrapper that can be put in the fire pit and, once the paper is lit, the log easily catches fire.

Making a Place for Your Pit

FIRE PIT AND ROUND GRAVEL PATIO

A dedicated and visually pleasing space for fires can be created in your backyard using a section of steel culvert pipe for the fire pit. This is lined with concrete retaining wall blocks, while gravel topped with decorative rock forms a surrounding patio.

Step 1: As described in the previous fire pit project, determine the perimeter of your fire pit using a stake and a string. Mark the circular boundary with spray paint. Make sure the diameter of your pit matches the diameter of culvert pipe you plan to use. A section of pipe that is 36 inches in diameter and 18–20 inches tall should be adequate for most backyard fires. Then use a longer string with the stake still in the center of the pit to create a larger circular boundary for your patio. Keep in mind that you will want room for seating and a place for firewood. Mark the patio boundary with spray paint.

Step 2: Remove sod from the patio area and grade dirt to make it as smooth as possible.

Step 3: Now you want to create a trench around the pit perimeter that will hold a gravel base and eventually hold the culvert pipe surrounded by retaining blocks. Using the string on the center stake, mark a circle that is six inches smaller than the existing fire pit radius. Then mark a circle that is larger than the pit radius and matching the width of the retaining wall blocks. Dig down four inches to create a trench and use a hand tamp to compact the soil at the bottom of the trench.

Step 4: Fill the bottom of the trench with compactable gravel and thoroughly tamp it down to form a hard surface that will be impervious to shifting and erosion. Make sure it is level. Then fill the entire circular patio area with compactable gravel that reaches to two inches below the surrounding lawn.

Step 5: Position the culvert pipe in the trench so that it is perfectly centered using the stake as a guide. Adjust gravel in places to make sure it is level.

Step 6: Lay the first layer of retaining wall blocks in the trench and placed flush around the culvert pipe. Make sure the blocks are level on top.

Step 7: Place additional layers of retaining blocks to create a wall that rises slightly above the top of the culvert pipe. Stagger vertical joints and, once they are level, seal in place with masonry adhesive.

Step 8: Using a plate compactor, harden the gravel throughout the patio area to create a solid sub-base. Install edging around the perimeter of the circular patio.

Step 9: Add a two-inch layer of decorative, angular rock (not rounded pebbles) that lends itself to being tamped down. Compact the rock with a plate compactor.

CIRCULAR PAVER PATIO WITH FIRE PIT

Compared to bricks, which require being mortared in place, concrete pavers are an easy DIY way to install a backyard patio. The uniform blocks come in all dimensions and colors and can be placed over a simple sub-layer of packed sand and gravel. The pavers are often sold as "kits" to create a specific size patio. A circular patio with a fire pit in the center is easy to install with a circular paver kit. To accommodate adding a fire pit, modify the factory design of any circular kit. Instead of installing the paver centerpiece block according to kit instructions, place a fire pit in the center following the steps described in the "Gravel Patio" instructions on pages 74–75. Once you have installed the steel culvert pipe for the pit, then you can fit the circular paving stones around it and build a wall of matching pavers to encircle the pipe. Or you can buy a circular paver patio kit that includes a fire pit as part of the design.

To Burn or Not to Burn

There are few places in the United States these days where the environment is moist and stable enough to safely have a wood-burning fire any time of year. Investigating local restrictions should be the first line of action in deciding whether or not to have a fire in your backyard. A call to the municipal fire department or a search on the city's website should produce the information you need. However, if you are in a rural area that is unregulated or you simply can't locate information about whether or not it is OK to have a fire on a particular day, here are some guidelines:

RED FLAG WARNING
This term is used by the National Weather Service when fire weather forecasters determine there are "extreme burning conditions" in a particular geographic region. The criteria for a red flag warning involves an area that has experienced a dry spell of a week or more, or the conditions occur during a dry time of year such as fall. The other parameters are: a sustained wind speed of 15 mph or greater; a relative humidity of 25 percent or less, and a temperature

greater than 75°F. Wood burning fires should not take place on red flag days. And even natural gas or propane fires can be unruly in extreme wind with flames going out or licking items nearby.

NATIONAL FIRE DANGER RATING SYSTEM
This set of indices used by fire managers on all U.S. Forest Service lands is an assessment of that day and the next day's fire danger in specific regions. While these ratings help land managers with recreation planning and anticipating fire hazards, the system is also handy for recreationists on or near the public lands. The rating is likely familiar to anyone who has camped in a Forest Service campground and seen the Smokey Bear sign displaying the fire danger level for that day. An interactive map giving the daily ratings can be found on the U.S. Forest Service Wildland Fire Assessment System website.

Follow this guide for interpreting the Forest Service fire ratings:

- *Low* – Fuels do not ignite easily from small embers but a more intense heat source, such as lightning, may start fires.
- *Moderate* – Fires can start from most accidental causes. Most wood fires will spread slowly to moderately, although they will burn quickly through dry grasses. Fires are not likely to become serious and are easy to control.
- *High* – Fires can start easily and unattended campfires and brush fires are likely to escape out of control. Fires can become serious and

FIRE KEEPER: MARK SHIERY

Mark Shiery says he first became "smitten with the fire bug" when he was getting a degree in Forestry from the University of Wisconsin in the early 1980s. Since then he has worked with fire in just about every capacity—as a hotshot fighting wildland fires for the U.S. Forest Service and as a structure firefighter for the City of Flagstaff. And somewhere in between Shiery worked as a logger.

Early in his career, Shiery traveled all over the West fighting fires on public lands as part of hotshot crews. Wildland fire fighting back then was like guerilla warfare with little technology for communication or other support. "It was not the sophisticated operation that it is today," laughs Shiery. "We would just get dropped off and work as long as it took to put out the fire. We didn't get time off and there were no supplies coming in."

Shiery spent two months in 1988 working on the fire in Yellowstone National Park. Nearly eight hundred thousand acres burned and a crew of more than nine thousand firefighters battled the blaze. "The fire burned so hot that I could see columns of gas being released from the trees and rising up before the fire reached them," recalls Shiery. As a municipal firefighter Shiery helped establish a wildland fire program for the city of Flagstaff. Like many rural areas in the West, Flagstaff bumps up against large stretches of national forest and is at significant risk for wildfires within the city limits.

The program Shiery helped create uses the same principles for minimizing wildfire on public lands but applies them to residential areas in the city. In addition to banning uncovered, wood burning fires in backyards, the program encourages residents to clear their yards of dry pine needles and debris and also promotes thinning of forests inside the city limits.

Shiery is now retired but still consults for the U.S. Forest Service and other agencies on fire safety and management. He especially enjoys camping and sitting around a campfire in the forest. And his extensive professional fire experience helps him appreciate things that the average camper might not.

"When my buddies and I find a fire ring that is full of burned wood, we are delighted because we know that wood will light and burn easily to get our fire going," he says. "We call it wood that has experience."

spread rapidly unless they are put out while they are still small.

- *Very high* – Small fires can quickly become large fires and exhibit extreme fire intensity, such as long-distance spotting. These fires will often become much larger and longer lasting wildfires.
- *Extreme* – Fires of all types start quickly and burn intensely. Spot fires are probable making these fires very difficult and dangerous to fight.

THINK ABOUT THE TRIANGLE

"When considering whether or not to have a fire in your backyard, you want to consider three critical factors influencing fire behavior: weather, fuels, and topography," says retired firefighter Mark Shiery. These factors determine what will happen after a fire has started, especially if it gets out of control. The three criteria provide a gauge used by U.S. Forest Service managers as well as fire weather forecasters for the National Weather Service. Weather that is conducive to wildfire involves high temperatures, low humidity, and wind—in any combination. But weather alone will not cause a wildfire and that is where the fuel level comes in. The more fuel that is dry and closely spaced raises the threat factor. In the forest, the fuel comes from trees and the understory. But in residential areas homes are the fuel. Topography can increase the threat when fire happens in hilly or steeply sloped areas that give the flames a runway for picking up speed.

"You have to think about how are all these factors going to combine together," notes Shiery. "And the biggest concern is wind. If all of a sudden, wind takes over, then the fire is out of control and you can't stop it. The embers are flying and can land far-away in places you can't see."

AVOID A BONFIRE

Regardless of the fire triangle conditions, a large fire is hard to control as well as safely extinguish. Flagstaff, Arizona, municipal firefighter Jeff Bierer recommends using no more firewood at one time than can fit in a five-gallon bucket. Having the bucket near the fire pit to use as a measure can help prevent the constant temptation of throwing another log on the fire just out of habit or boredom.

Shiery also offers an additional way to keep things under control: "If people are backing up, then you have too much fuel on the fire."

Also, always have a bucket of water and a water hose nearby in case embers cause spot fires in your yard. And once the party is over, the process for putting out a backyard fire is the same as for extinguishing a campfire described in Chapter 3.

CHAPTER 5

Keeping Cozy Indoors

Whether you are in need of warming up a remote mountain cabin or interested in adding a woodstove for extra heat in your own home, the more knowledge you have, the better your BTUs

---

During the decade I lived with only a wood stove for heat, I had one enduring dream. As I angrily swung an axe to split wood under driving snow or tried to coax the stove to life on freezing mornings or endlessly shoveled ashes, I vowed that one day I would move into a home with central heat. There would no longer be sweat equity or countless hours of fire tending required just so I could be comfortable in my house. When it was cold, all I would need to do would be to turn up the thermostat. Such convenience seemed almost decadent during those days.

HOME IS WHERE THE *HYGGE* IS

According to the Collins English dictionary, *hygge* as a noun refers to "a concept originating in Denmark, of creating cozy and convivial atmospheres that promote wellbeing." In 2016 the Collins dictionary named *hygge* (pronounced *HOO-gah*) the runner-up for "word of the year" after Brexit.

While *hygge* has been central to Scandinavian life for centuries or longer, it recently caught on in the UK and the United States. In 2016, more than 20 how-to *hygge* books were published in English, with the bible being *The Little Book of Hygge* by Meik Wiking. Not coincidentally, Wiking is also the founder and chief executive of the Happiness Institute, a Copenhagen-based think tank that researches why some societies are happier than others. Denmark—the *hygge* world capital—regularly rates No. 1 in United Nations surveys for the happiest country. Wiking contends that happiness and *hygge* are linked. This seems counterintuitive since the winter weather in Denmark is exceptionally dreary, so it would make more sense that its citizens should be stricken by seasonal affective disorder (SAD) instead of filled with joy. But well-done *hygge* is obviously very effective in fighting off the wintertime blues.

Hygge-ness can be experienced either alone or with loved ones, but preferably while you are tucked under a wool blanket in a *hyggekrog* (Danish for "cozy nook"). A warm fire is essential as are numerous lit candles. *Hygge* cuisine includes plenty of hot tea, cake, and especially a steaming cup of glogg (mulled wine) and muesli served in a wooden bowl. For more tips (and products), check out www.hyggelife.com.

Eventually, my dream came true. I bought a home with central heating powered by a natural gas furnace. Certain I would never again tend an indoor fire, I gave away my ash bucket, wood-splitting axe, and fire tools. When a heavy snowstorm was forecast, I no longer had to prepare for it by splitting extra wood and hauling it into the house. My mornings were free from futzing with warm coals I had banked the night before to make fire starting easier. It was complete, automated bliss—at least for the first few years.

The angled walls of a Rumford fireplace send heat out into the room.

After the novelty of the convenient thermostat wore off, I started to miss the wood stove. My house seemed sterile without the crackling pop of a fire. And while being able to keep my house at one temperature was wonderful, it was not cheap. I found myself turning the thermostat down on frigid nights to save money and I longed for the intense warmth I could generate—at zero cost—simply by putting another log or two in the stove. Without the wood stove, winters in my centrally-heated house were significantly less cozy feeling. The Danish have a specific term for this desirable coziness: *hygge* (see sidebar above). It has to do with a certain aesthetic of wholesome, comforting warmth that can be especially enjoyed during an unforgiving winter, and a fire is central to achieving optimal *hygge*.

Soon enough, I was shopping for a wood stove and hired a mason to tile a corner of my living room so I could safely install the heater and chimney. Compared to the decades-old stove in my previous house, the efficiency of modern stoves has improved greatly and I was able to buy a model that was not only warmer for the size than what I had before but also cleaner burning and easier to operate. Maybe it was because I now had the option of not hassling with it and just turning up the thermostat on my gas furnace, but I actually had a renewed enthusiasm for all that went into using the stove. While it required work, there was a real payoff for my efforts in terms of significantly upping my home's *hygge* factor.

"We have become so used to the conveniences of modern technologies that we have lost our appreciation for the complexities of heating a home with wood," says primitive technology expert Don Kevilus. As the owner and founder of Four Dog Stove Company, Kevilus designs and makes wood stoves for use in tents and has a deep affinity for wood heat. "In the United States, 200 years ago, most people were heating their homes and cooking with wood. From the time a child was able to walk, they were accumulating the life-long skills necessary to harvest wood, maintain a fire, and cook with fire."

Kevilus says many people in first world countries have lost these skills over the span of two or three generations but getting them back just takes practice. "The more you do, the more you know, and the better you get," he says.

FIRE AND PLACE

History of the Hearth

The earliest evidence of human ancestors using an established fireplace dates back three hundred thousand years to a site called Qesem Cave in Israel. The "hearth" identified by archaeologists was located in the center of the cave. The excavated soil showed long term use with layers of ash, along with pieces of burnt animal bone and flint tools used for carving meat.

Microscopic analysis of the ash in the hearth indicated that the fireplace was used over and over, perhaps daily, for years. Tools were being made around the flames. Food was cooked and shared with the group. And presumably the human ancestors made it their communal gathering place. The fire site was likely the nexus for an emerging domesticity where the hearth was literally and figuratively the heart of the home. As proof of how much cooking happened there, the bones of 4,740 prey animals were excavated from the cave.

By the time humans moved from caves into primitive built structures, our ancestors started perfecting the art of burning a fire indoors. Keeping

heat inside while sending smoke out became an ongoing challenge. One of the most ingenious and long-used methods came from the indigenous people of the North American plains. The portable cone-shaped structure made of animal hide commonly called a teepee had a fireplace in the center of the circular floor with smoke venting directly above through the small opening at the peak of the cone.

An adjustable flap at the top along with the door flap at the bottom of the teepee made it possible to regulate airflow and create a vacuum to pull smoke up through a kind of chimney. Once the fire got going and was producing less smoke, the flaps could be closed to hold the heat in.

In other primitive houses made of stone or wood, the fireplace was located against the struc-

ture's exterior wall so smoke could go out a hole in the wall above the flames. Even though more sophisticated venting technology existed in ancient Rome—including bakeries with flue pipes—that knowledge was seemingly lost for a thousand years after the fall of the Roman Empire. During the early Middle Ages in Europe a metal canopy or hood was added above the fireplace to direct smoke out the wall hole. Eventually came the idea of recessing the fireplace into the wall to further contain the smoke by having a barrier above and on three sides.

Sometime in the twelfth century, the chimney with a flue was invented in Europe, which finally provided an effective way to vent smoke and fumes and improve indoor air quality. The concept behind the flue, which is the interior chamber of the chimney, revolves around the simple fact that hot air rises. And the taller and straighter the flue, the greater the suction created as warm air and smoke in the fireplace is drawn toward the cold air outside at the chimney top. As the plume of hot air and smoke travels up the flue, cooler ambient air in the room is drawn to the fireplace, feeding the fire with oxygen. When the chimney was located in the center of larger houses, it contained multiple flues for fireplaces in different rooms. In addition to the flames themselves, the radiant heat emanating from the masonry chimney also provided warmth even for a few hours after the fire had gone out.

The Great Hall of Edinburgh Castle in Scotland, completed in the sixteenth century, features an elaborate fireplace.

European cities became a sea of towering chimneys, and this gave the various ruling monarchies an idea. In the 1340s, King Phillip of France imposed the first "hearth tax" that would become a much-despised burden waged by monarchies across Europe. Rather than taxing the number of heads in a family—since people are easy to hide—every chimney was taxed. In 1662 the English Parliament also imposed a hearth tax, which became known as "Peter's pence," to support the Royal Household of King Charles II. While peasants went to great lengths to try and hide their hearths from the tax collector, stuffing chimneys with hay and smoking up their houses or even connecting through a wall to a neighbor's chimney, the aristocracy of seventeenth-century Europe flaunted their wealth with elaborate fireplaces. English Tudor fireplaces were adorned with ornamental over-mantels sporting the family crest carved into wood panels. Meanwhile French Baroque castles contained flamboyant fireplaces with mantels carved in marble and framed by columns and caryatids.

When the Puritans arrived in North America in the seventeenth century—in part to escape the King's hearth tax—they took the fireplace to a whole new level. In order to survive the sub-zero temperatures of New England winters the Pilgrims built small cabins with giant fireplaces. The Colonial hearth was typically four or five feet tall and at least ten feet wide and, unlike those built by Europe's wealthy aristocracy, these fireplaces were all about utility. There was a recessed shelf in the back for baking and a move-

This American Colonial fireplace was designed for cooking.

able crane where cast iron pots and teakettles could be suspended over the flame. Utensils hung from hooks on the mantel and on especially frigid nights the family stayed warm by literally sitting inside the fireplace. But cold air came barreling down the chimneys of these giant hearths when not in use, so a fire had to be kept burning around the clock. Fortunately for the early European settlers, New England forests were flush with firewood.

However, a few decades before America's founding fathers were drafting the Constitution, the colonists' giant fireplaces were no longer sustainable. In little more than a century, the forests of New England had been cut over by homesteaders. The dearth of firewood was especially problematic in growing cities like Boston and Philadelphia. Ben Franklin—ever the inventor—decided to tackle the issue and create a more efficient method for home heating. He sought to solve the problem of so much warm air escaping up the chimney and cold air sneaking down the flue. Additionally, Franklin saw that uneven airflow coming into the hearth from an open room was leading to incomplete combustion with lots of smoke and half-burnt logs. He reasoned that if he could contain the fire it would hold the heat in to burn fuel more completely while also preventing the escape of warm room air up the chimney.

In 1740 Franklin created what he called the "Pennsylvanian fireplace." This iron box nestled inside the drafty brick fireplace and had a partial door on the front to control room airflow and ensure more complete combustion of wood. By having the fire nearly contained within the box, radiant heat could warm the room without allowing cold air to enter the room from flue down drafts. The pumping action of a fire in an open fireplace has the tendency to draw far more air from the room than it needs for oxygen supply. Franklin solved this problem on his fireplace with the addition of the door and a vent underneath the box that drew less air but enough to keep the fire going. Additionally, Franklin added a moveable iron flap inside the chimney—called a damper—that could further regulate air in the flue. The damper could be closed when the fireplace was not in use or partially closed

Old-fashioned potbelly stove

once a fire was going to let smoke and fumes out but not allow so much warm air to travel up the chimney.

Modifications to Franklin's fireplace led to the Franklin stove, which significantly improved heating

efficiency compared to the fireplace. Often located in the middle of the room, these cast iron potbellied stoves were fully enclosed and able to provide more warmth than a fireplace for far less fuel. For most of the nineteenth century potbellied stoves were the heating method of choice in households across the United States and Europe. Growing cities during the industrial revolution switched to coal over wood as a heating source in potbellied stoves, which was easier to transport to urban areas and increasingly plentiful as coal deposits were discovered in Pennsylvania and the rest of Appalachia.

However, the fireplace was not dead and continued to be used in homes as well, even if mainly for reasons of *hygge* rather than heating. The potbellied stove, which largely hid the open flame, was not nearly as aesthetically pleasing as the living room hearth. Fireplace efficiency was also improved in 1795 when an English man named Benjamin Thompson—a.k.a. Count Rumford—introduced a new design that became called the Rumford fireplace. Rumford flipped the rectangle of the firebox so that the fireplace was far taller than it was wide. In addition, he angled the sidewalls of the firebox and narrowed the opening (called the throat) to the chimney by slanting the back wall of the hearth toward the room. This design greatly increased the radiant heat coming into a room from the fireplace and it also minimized the amount of warm air being sucked up the chimney. Most homes built after 1800 follow the Rumford fireplace design, including modern masonry fireplaces.

WOOD HEAT 2.0

Anatomy of a Fireplace

With the exception of alternative fireplaces (see Home Heating Alternatives on pages 108–11), the technology of the hearth has remained relatively unchanged for the last 225 years. What Count Rumford realized and remains key to a vigorously burning, smoke-free fire is to keep heat circulating in the firebox so that it warms the flue and channels smoke up the chimney with a strong draft—not to mention radiates warm air out into the room. The minute the flue and the firebox start to cool, not only does the wood burn less efficiently, but the smoke sinks rather than rises, often making its way into the room.

The three walls and floor of the modern fireplace are made of special firebricks, which are porous and retain heat far longer than regular bricks or other masonry. Directly above the fireplace is the throat opening to the chimney. In order to create a strong draft up the flue and prevent warm air from escaping the fireplace, the width of the throat must be narrower than the flue. Located inside the chimney and often extending several feet above it, the flue is a smooth metal chamber that provides an efficient path for smoke and fumes to escape. The flue should be at least several feet taller than the highest point on a roof's pitch in order to have optimal venting. Many mod-

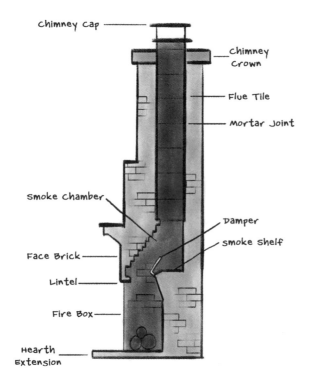

Chimney Cap

Chimney Crown

Flue Tile

Mortar Joint

Smoke Chamber

Damper

Smoke Shelf

Face Brick

Lintel

Fire Box

Hearth Extension

When the fireplace is not in use the flue, should be closed completely.

On the house exterior, the top of the masonry chimney should have a beveled concrete lid that sheds and shelters the chimney from debris. The flue rising above the chimney should be sheltered with an umbrella-like cap and metal screen that allows smoke to escape but keeps birds and critters from getting inside. Metal flashings are installed around the area where the chimney meets the roof to protect from water seepage and erosion of the masonry. While most modern chimneys are lined with metal or masonry that are designed to withstand extreme heat, it is important to have a chimney cleaned once a year (see pages 106–7) and inspected for any cracks or leaks. The safest bet is to use a chimney sweep company certified through the Chimney Safety Institute of America (www.csia.org). The organization's website has a directory of certified companies that can be located by zip code.

As for fireplace accessories, the single most

ern fireplaces also have a smoke shelf or curved chamber at the bottom of the flue designed to capture smoke that sinks when the fireplace cools before it infiltrates the firebox. A damper is positioned either at the bottom of the flue or near the top, following the same design that Ben Franklin invented. This hinged metal door has an adjustment knob on the hearth and controls airflow in the flue. It should be fully opened when a fire is getting started to allow maximum air circulation and heating of the flue. The damper can be partially closed once a fire stops smoking in order to restrict cold air from coming down the chimney.

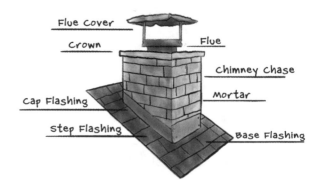

Flue Cover

Crown

Flue

Chimney Chase

Cap Flashing

Mortar

Step Flashing

Base Flashing

important item is a screen or doors that completely wrap around the hearth to block embers from getting into the living space. Fireplace users should be vigilant about properly replacing the screen after futzing with the fire. A more secure option is doors that fit flush against the hearth front. Glass doors offer the added benefit of keeping warm room air from disappearing up the chimney.

Even when it seems like the fire is dying down, sap deep inside a log can cause wood to pop apart and send embers flying several feet into the air where they can land on upholstery or paper and start a house fire within seconds. In addition to the screen, it is important to have a large area of stone in front of the hearth or a fireproof hearth rug that covers carpet or wood floors. Along with a poker for moving logs around, hearth gloves are handy for more delicate repositioning of smoldering logs or protecting hands when using the poker over open flame. Fireproof insulated suede gloves offer a solid grip and full protection. You also need a sturdy iron grate that keeps logs at least several inches off the ground and secures them against shifting or collapsing as they burn.

Anatomy of a Wood Stove

The design and technology of wood stoves has evolved considerably since the old potbelly models used in the 1800s. However, those first stoves and modern wood heaters still operate on the same basic principles: an air intake underneath the stove

YOU KNOW YOUR FIRE-PLACE IS PERFORMING PROPERLY IF . . .

- Fires are easy to light and a draft up the chimney builds quickly.
- When you light newspaper as kindling, the smoke immediately moves upward.
- The bricks in the firebox are tan in color after a fire and not black.
- There are no foul odors or cold air coming from the chimney when the fireplace is not in use and the damper is closed.
- The exhaust coming from the top of the chimney is clear or white. A plume of blue or grey smoke indicates poor combustion.

controls oxygen flow to feed the fire inside a fully enclosed firebox. A long, thin stovepipe pulls smoke and fumes up and out, while dampers allow for adjusting the draft to achieve a more complete combustion of fuels. Unlike a fireplace where a person must sit close to open flames to feel warmth, the wood stove heats a much larger area through its radiant properties. And no electricity is required.

Most wood stoves are made either with the tried and true cast iron or newer welded steel. Cast iron is considered more durable, especially with larger stoves, and it is more aesthetically pleasing, often with curved legs and old-fashioned designs. Colored enamel is also sometimes added to a cast iron stove to improve its appearance and make its exterior easier to clean. Steel stoves are usually made in no-frills designs but come at a lower price than the cast iron varieties.

In 1988, the U.S. Environmental Protection Agency implemented guidelines for wood-burning stoves that required all models sold in 1990 or later meet air quality emissions standards. These rules not only significantly reduced air pollution from wood smoke but also improved the heating efficiency of stoves. EPA-certified stoves generate at least one third more heat for the same amount of fuel as pre-1990 models. EPA-approved "phase II" stoves must produce less than 7.5 grams of smoke per hour, while older stoves release 15–30 grams of smoke per hour into the atmosphere. In 2015, even stricter EPA performance standards were released that require all stoves sold by 2020 to emit no more than 4.5 grams of smoke per hour. Many stoves already meet or exceed this new standard.

Reducing emissions from wood smoke is not only important for controlling air pollution that is harmful to the environment but particulates from wood smoke are also dangerous to human health. Tiny airborne particles from burned wood can lodge in the lungs and cause a variety of ailments including asthma, chronic bronchitis, and cancer. And because these new stoves burn more efficiently, EPA certified models also reduce indoor air pollution by minimizing ash and stray smoke. While buying a 50-year old wood stove off

Craigslist that is fully functional and in good condition may seem like a bargain, there are potentially high costs in terms of health.

EPA certified stoves come in two varieties: catalytic and non-catalytic. Both technologies reduce emissions by heating residual smoke and fumes at the top of the firebox in order to completely burn particulates before the air exits up the stovepipe. In catalytic stoves, the exhaust is passed through a honeycomb-like ceramic piece coated with a catalyst—usually palladium or platinum—that ignites the unburned particles in smoke. Meanwhile non-catalytic stoves achieve a comparable clean burn by using a combination of firebox insulation, a baffle to produce a longer and hotter gas flow path from the burning wood, and a system to introduce pre-heated combustion air near the top of the firebox. Catalytic stoves are typically slightly more expensive and require vigilant maintenance to replace the worn catalyst, but they also provide longer burn times and maintain a more even heat. Non-catalytic stoves are often more aesthetically pleasing with a larger glass door to view the flames but the fire is more erratic and maintaining it requires a bit of futzing.

Another measure for modern wood stoves is heat output, which is quantified in BTUs. The BTU number refers to the stove's combustion efficiency in terms of the percentage of total heat content in the wood fuel that can be extracted when burned as energy. Beyond meeting the EPA efficiency standards, the BTU output of a stove is directly related

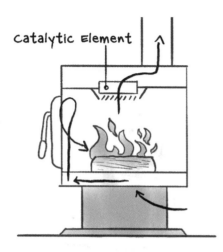

Catalytic wood stove

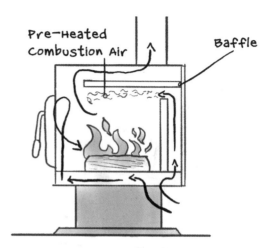

Non-catalytic wood stove

to its size. The bigger the stove, the more wood you can put in it to generate heat and the larger the surface area of the firebox for radiating that heat. A general rule is that 2.6 square feet of firebox bottom is required to heat each 1,000 square feet of room area. However, this guideline only applies to large open areas rather than multiple rooms. It is more effective in terms of avoiding drafts to have multiple smaller stoves in different areas of a house rather than one giant stove in a central location.

As for the stove chimney, most city fire codes in the United States require double-walled insulated pipe as a minimum but many modern stove installations come with triple-walled stovepipes. This reduces the chance of intense heat or flames being transmitted to surrounding structure materials and causing a house fire. Like a fireplace chimney, the stovepipe must also be several feet taller than the highest point of a roof.

Some wood stoves have air circulators built into the firebox that help blow warm air out into a room. But a ceiling fan strategically placed above the stove also helps disperse heat. As the warm air rises, the fan sends it back down into the living space. The intense radiant heat coming from the stove also dries out indoor air and can become uncomfortable without some kind of humidifier. This can be as simple as keeping a full teakettle on the stove top, which is also handy for having a constant supply of hot water. Cleaning the stove is also critical to keeping it in prime working condition and avoiding chimney fires (see pages 106–7).

Modern Wood Stove

KEEP THE HOME FIRES BURNING

Never Skip the Basics

Successfully starting and maintaining a hot, clean burning fire in a fireplace or wood stove relies on the same four-step fire fuel ladder principle described in Chapter 3 for campfires. Ignition should progress from smaller to larger fuels. Step one is tinder such as newspaper; step two is pencil-sized sticks; step three is finger sized sticks; and, finally, step four is the larger fuel logs.

"The main thing people do wrong when heating with wood is they don't do enough preparation before starting the fire," says wood stove maker Don Kevilus. "Just five minutes of extra prep will allow you to have a hot fire going within 10 minutes. But if you don't invest the time beforehand preparing the fuels, it will take you 15–20 minutes to get a slow burning, punky fire going."

Kevilus advises keeping fuel dried and sorted well in advance of fire starting. Tinder such as grasses, tree bark, and pinecones can be gathered during hikes or dog walks and kept in a breathable basket indoors but well away from the fire. Newspaper (including in the form of knots; see page 106) is excellent kindling and a good reason to continue subscribing to old-fashioned print editions of the daily news. Kindling should be plentiful and sorted in a dedicated container. Kevilus likes to use the commercially available Fatwood sticks. It is also easy to gather this size wood when splitting logs in the backyard. Rather than leaving the wood splinters on the ground, pick them up and put them in your kindling box. Fuel logs should be split and dried. Part of your daily fuel preparation should be to bring in wood for the next day so it has time to dry indoors, preferably a few feet away from the stove. This will minimize smoke and reduce the amount of heat energy wasted on drying the wood before it can reach full combustion.

"If you are using dry wood and still getting a lot of smoke, you have probably put on too much wood too soon," says Kevilus. "You need to get a bed of coals going before putting the larger logs on the fire."

As for what type of wood to use, resin-infused varieties such as pine (a.k.a. Fatwood) and spruce

TRY THIS: MAKE A WOOD CARRIER

Rather than risk getting splinters and covering your jacket with bark by cradling wood in your arms, follow these steps to make a carrier that is convenient for use between the house and the woodpile.

1. Find a piece of heavy canvas or other sturdy cloth. Cut it to be about 13 1/2 inches wide and 50 inches long. Hem the edges.
2. Get two 1-inch diameter dowel rods and cut them to be about an inch longer than the hemmed cloth is wide.
3. Get two 15-inch lengths of rope or heavy nylon cord.
4. Drill two holes in each dowel rod that are about 3 inches on either side of the rod's mid point. The hole should be just big enough for the rope to pass through.
5. Lay one of the rods across one end of the canvas and fold the material over it so it can be sewn to snugly enclose the rod (but don't sew yet). Punch holes in the canvas where it covers the holes in the rod.
6. Tie a knot in one end of the ropes. Pass the other end through one of the holes in the stick and pull it tight.
7. Pass the rope through the matching hole in the canvas.
8. Pass the rope through a 4 3/4 inch length of garden hose or flexible tubing to make a comfortable handle.
9. Pass the rope through the other hole in the canvas and then through the other hole in the rod and tie a knot in the end.
10. Turn under a 1/2 inch hem and sew the canvas together enclosing the stick and the handle knots.
11. Repeat steps 4–10 to make the other handle and the carrier is finished.
12. To use the carrier, lay it on the ground, pile wood on one half of the canvas. Grasp both handles and pick up the load. Bounce the wood on the ground once so that it settles. Tuck an extra log or two in the top and carry your load to the house.

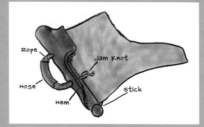

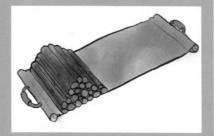

are excellent for kindling and getting the fire started. However, these and other softwood species should be avoided as a primary fuel because the resin causes creosote to build up in chimneys. This can lead to chimney fires if the chimney is not regularly cleaned. Hardwoods such as oak and maple are ideal for fuel logs and offer the most heat in terms of BTUs (see chart in Chapter 6).

The Grate Debate: Two Ways to Build a Fire in a Fireplace

Even though fireplaces have been in use for as long as humans have lived in houses, it has only been in the last thirty years that a new approach to fireplace burning has taken hold. The age-old method involves using andirons and sometimes a grate to hold logs. The fire is built gradually, starting with kindling and then adding larger fuel pieces on top. The second more recent method involves stacking fuel directly on the firebox floor with the large logs on the bottom and ascending to the kindling on top to create what is called "top-down" construction. There are advantages and disadvantages to both approaches. Deciding which method is best for you depends on the condition of your fuel and the kind of fire you are seeking.

THE ANDIRON APPROACH

Andirons are L-shaped cast iron implements that come in pairs and have been used for centuries to create a barrier against burning logs escaping the fireplace. They also add decorative flare to the fireplace and are made in many styles. While logs can be balanced on the two legs at the base of andirons to provide airflow underneath fuel logs, a basket-like iron grate may be placed in between the andirons to further improve air circulation underneath large logs to promote the combustion process. Historically, andirons were used when the fireplace was the home's only source of heat, and cooking and large amounts of wood—which was often green and not split—was burned. The increased airflow around the logs dried out the logs and the andirons kept the un-split wood from rolling onto the floor.

Just like our pioneer ancestors, building a fire using andirons and/or a grate is still best if you have wood that is green or not split and needs that extra airflow to readily burn. You may also choose this method if you desire a bright fire for a short period—such as having a glass of wine after dinner in a cozy living room. If the fireplace is drawing properly, the airflow moving underneath split, dried wood quickly ignites logs for a vigorous burn that gives off a blast of light and heat.

Before starting the real fire, light a flash fire with a small pile of kindling simply for the purposes of warming the flue and improving its updraft. Then place several small split logs (about forearm size) in the grate. Keep about a half-inch of space between the logs to allow air circulation. Stuff kindling and newspaper underneath the grate but not packed so tightly that airflow is impaired. Place some kindling and tinder on top of the fuel logs as well. Light the

tinder and blow periodically to encourage flames to move up to the kindling on top.

Getting a fire started with this method typically requires three stages of fuel after the flue has been warmed. First, light kindling and smaller (forearm size) fuel pieces as described above. After about 15 minutes, add medium sized logs. And, finally, place the larger logs on top of the blazing fuel stack. As the fire dies down, add kindling to the top of partially burned logs to stoke the blaze before throwing on another fuel log.

A TOP-DOWN FIRE

Although this method has been used for centuries in large, industrial masonry heaters, it has only recently become popular for home fires. The top-down construction is especially useful in wood-stoves due to its convenience and prolonged heating power but it also performs well in fireplaces for all the same reasons. The principles at work with a top-down fire in a fireplace are the opposite of the andiron fire. Instead of the ignition force building from underneath the grate with air circulation, the top-down construction begins with kindling and tinder igniting at the peak of a fuel pyramid and the fire gradually working its way to the base logs on the fireplace floor.

A top-down fire is appealing primarily for reasons of convenience. Once the fire starts, it generally feeds itself by consuming progressively larger pieces of fuel as the flames move downward. And because the kindling is on top of the pile, it immediately warms the flue when lit and negates the need for a flash fire before starting the real fire as described above. A top-down fire is also longer lasting because the logs don't burn as rapidly due to less airflow. However, the fire is also not as bright and the pile of wood on the fireplace floor is not as aesthetically pleasing as glowing logs on stately andirons.

To build a top-down fire, remove the grate from the firebox. You may also need to remove the andirons if they prevent fuel logs from resting flat on the fireplace floor. Because you are essentially constructing a tower of fuel, it is important that the base is stable and sitting directly on the firebricks. Another essential condition for a good top-down fire is wood that is split and dried. Otherwise, the fire will be smoky and un-split logs will roll and shift.

Once you lay your largest fuel logs side by side to form the base, add 4–5 layers of progressively smaller pieces of fuel; each layer should be placed perpendicular to the one below it. Build the pyramid to a size that is no more than half the height of the firebox. Finally, top it with kindling and then tinder. See the instructions below for "building the perfect top-down fire." The process in fireplaces is the same as that used in wood stoves.

"I had always built my fires the traditional way," says Jason Brown, a certified fireplace technician in Flagstaff, Arizona, who has been servicing fire-places for more than a decade. "When I tried my first top-down fire I was amazed at how well it worked. I didn't have to do anything after I lit the kindling and there was hardly any smoke. It's funny that we are just now figuring this one out."

Wood Stove Fires

When it comes to lighting a fire in a wood stove, there are different philosophies about which strategy is best but the one that seems to have the most adamant number of followers is the "top-down" method. "Since the early 1990s, it has become de rigueur among many in the [wood heating] business, and by extension, their customers to build their fires top down," writes John Gulland on the Canadian informational website Wood Heat (www.woodheat.org). The benefits top-down enthusiasts report (including myself using my new wood stove) include: minimal start-up smoke; no chance the fire structure will collapse and require starting over; just the right amount of air reaches the fuel, and there is no need to open the door after loading to add larger pieces of kindling which reduces hassle and indoor air pollution.

Feeding the Beast

MAINTAINING THE STOVE FIRE
While it is good practice to keep adding a log to a fire in a fireplace as the fuel burns down, this is not

TO BUILD THE PERFECT TOP-DOWN FIRE, FOLLOW THESE STEPS:

1. Remove ashes from the firebox so plenty of air can circulate around the fuel.
2. Place 3–4 large, split fuel logs laying side by side about one inch apart in the bottom of the firebox. Fuel logs should never be so big that they touch the walls of the firebox.
3. Place a layer of smaller split wood pieces on top of the base and running perpendicular to the logs below.
4. Lay a layer of still smaller sticks on top and perpendicular to the wood below.
5. Place a layer of fine kindling pieces in cross hatch fashion on top of the sticks.
6. Crown the fuel pyramid with a few sheets of wadded-up newspaper (or tied into knots; see sidebar on page 106).
7. Light the newspaper and allow as much air as possible. Once the fire has established a coal base you can dial down the dampers.

Note: When building a top-down fire in a fireplace, the fuel should extend up to no more than half the height of the firebox. However, in the contained environment of a wood stove, it is safe for fuel to be within a few inches of the top of the firebox.

the best method with a wood stove. The goal with a stove is to avoid smoldering fuel and allow the efficiencies of the firebox design to achieve complete combustion. Instead of adding a log to the stove fire every hour or so, wait until all the fuel logs burn down to coals. Then rake the coals into a shallow pile at the front of the stove near where the air intake is. Place a few fuel logs on the coals and close the door. Keep in mind that the most efficient and effective wood stove fires last eight hours or less so avoid putting giant logs in the stove in an effort to keep the fire burning all day long.

NIGHT FIRES

A fireplace should never be left unattended overnight but a modern wood stove can be used to safely construct a long-burning fire. If your goal is to keep the fire going through the night, begin the process about an hour before retiring to bed. Let the fuel supply burn down to coals as described above. Then

rake the coals forward and place large fuel logs in the firebox with smaller pieces on top. Wait for the firebox to be full of flames and the exterior of the large fuel logs are covered with a layer of charcoal. The charcoal layer will insulate the interior wood and slows down the release of combustible gases. Then turn down the stove's air controls so that there is just enough air to keep a low-grade fire going (this will take some practice). In the morning you should wake up to a layer of ash banking some hot coals that can be fed with air and kindling to start the next day's fire.

PILES OF CHARCOAL

In very cold weather when wood stoves are burning all day, it is inevitable that along with the ashes there will be lumps of unburned fuel in the form of charcoal. Rather than sending this potential heat energy out in the ash bucket make use of it. If there is a large coal bed with pieces of char rake it all forward and place one large log on top of the pile. Open all vents to burn the log as fast as possible. This should consume the charcoal as well and, in the process, make your house extra toasty.

"Consider wood burning a challenge," says Kevilus. "Try to build your skills so that you can get to the point where you are getting the maximum amount of BTUs out of every piece of wood."

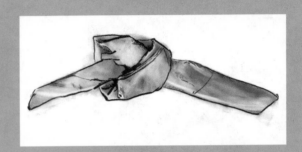

Maintenance and Safety

ASH REMOVAL

Cleaning a fireplace or stove must be part of the fire ritual. Ashes should be removed after every fire. This is important for getting proper air circulation around the fuel and it also is a healthier option. Shoveling large piles of ashes inevitably sends particles into the air of your living space. Use a metal container for holding the ashes since they can stay warm for days and cause fires when placed in trash bags or other flammable containers. Scoop out the ashes with a small shovel or a special ash trap, which is much cleaner. This wedge-shaped scoop has a sliding lid so ashes don't go flying around your house when you dump them in the can. Once you

have removed the ashes put them outside immediately, placed on a non-flammable surface like a cement driveway. Warm or hot ashes continue to smolder, even if it's not visible, and give off harmful carbon monoxide (see pages 107–8).

After the ashes have completely cooled they have a variety of uses. Ash is premium fertilizer in gardens and on lawns. Warm ash put on soil even protects seedlings against insects. Ash is also a great addition to compost piles for maintaining neutral acidity levels. And ash can be scattered on icy driveways and sidewalks for better traction.

CHIMNEY FIRES

Softwoods such as pine, fir, and spruce contain resins that build up inside a chimney, especially when the fire is smoldering. Try to minimize the amount you burn of these types of wood, using them mainly to get the fire started. Then switch to cleaner burning hardwoods. However, if you don't have access to hardwoods and are stuck with pine, have your chimney cleaned several times in a season. Otherwise, have your chimney or stovepipe cleaned once a year by a professional chimney sweep company. This is the only sure way to avoid the possibility of a chimney fire, which occurs when so much resin has built up inside the flue that the oil-like substance burns uncontrollably. This intense and sustained heat can cause surrounding materials in a house to catch fire and/or the constant spray of flames coming out the top like a blowtorch can catch the roof or trees on fire.

If you are renting a cabin and not sure if the wood stove or fireplace chimney has been properly maintained, here are some things you can do to ensure you will have a safe burn:

- Knock on the stovepipe. If you get a hollow tin sound it means the pipe is clear of soot. If there is a thud sound, it is either dirty or may have a rodent or bird's nest in it. Either way you should not have a fire.
- Build a small fire in the fireplace or stove to test its operations. If you get puffs of smoke coming out of the fireplace this means there are chimney problems. If you have a steady stream of smoke coming into the room it is more likely that you simply are not getting a good draft

up the chimney. Try opening a window to give more air to the fireplace, and also to vent the room. Also make sure the woodstove is not leaking smoke into the room through cracks in the metal.
- What color is the smoke coming out of the chimney? If it is white or clear, you have a good burn. If it is blue or gray, the fire is smoldering and not drawing well. Continuous black smoke means the chimney is dirty and you are at risk for a chimney fire.

CARBON MONOXIDE

When carbon based materials burn—whether it is in a fireplace, a gas furnace, or a camping stove in a tent—carbon monoxide is an inevitable byproduct of incomplete combustion. It is an invisible, tasteless, and odorless gas that can kill. Back in the nineteenth century when people lived in drafty cabins, carbon monoxide was not an issue because there was too much fresh air seeping into the living space. But now that homes are so well insulated, this deadly gas has nowhere to go and it kills unwitting victims every year. Early symptoms of carbon monoxide poisoning include nausea, vomiting, headache, and a loss of manual dexterity. If the poisoning progresses—often when people are asleep—the heart and lungs fail.

"Carbon monoxide will affect every person and animal in a home," says firefighter Jeff Bierer. "Everyone will have the symptoms of poisoning. If you suspect this, you should call 911 and open a window or door. Even if you suspect a stove might not be venting properly, crack a window just to be safe." Bierer says the best insurance policy against this threat is making sure you have a high quality dual function detector (that monitors both smoke and carbon monoxide) with batteries that are frequently replaced. If the alarm sounds, Bierer recommends calling 911 so the fire department can check the home to ensure it is safe. If a carbon monoxide or smoke detector is more than 10 years old, the sensor is prone to not working and the detector should be replaced.

FIRE EXTINGUISHERS

Another essential item for anyone burning wood in their home is the fire extinguisher. One extinguisher should be mounted on the wall between the stove or fireplace and the exit door. A second extinguisher should be on the wall in between the fire source and the main bedroom. Keep in mind that home fire extinguishers are sized for putting out small, trash-can size fires. If there is a fire, call 911 while also using the extinguisher.

HOME HEATING ALTERNATIVES

INSERTS

Exquisitely built stone or brick fireplaces in the living area of an old home can be the aesthetic soul of a house and one of its main draws for potential home buyers. But buyer beware. The fireplace was likely built back in a much draftier era when people did not blow piles of insulation in their attic, weather-strip every door, and install double pane windows throughout. Even if you repair the chimney to be in like-new condition, there is still the problem of a fireplace designed to feed off drafts in the living space. Without enough fresh air to feed it, the fireplace smokes and creates a frustrating dilemma of either opening a window and losing warm air or not using the fireplace to avoid smoking up the house.

A much better option than trying to use an old fireplace is to put a new, EPA-certified insert inside it. Similar to what Franklin designed with his Pennsylvanian Fireplace, an insert is fully enclosed like a wood stove. In fact, there is a wide selection of wood stoves that nestle inside fireplaces. But there are also inserts that appear more like a fireplace with an expansive firebox and a large glass front. Inserts vent through the chimney flue via a stovepipe, which should be installed by a technician certified through the Chimney Safety Institute of America. Unlike a wood stove in the middle of a room that radiates heat on all sides of the firebox, an insert sends that heat into the living space with a fan. The fireplace is lined with a metal box and the insert is placed inside that with several inches of space on all four sides for air circulation. The insert technology pulls in air from the room to feed the firebox while excess

Natural gas–fueled fireplace

FIRE KEEPER: DON KEVILUS

"I have been playing with fire since I was five years old," says Don Kevilus.

A native of Minnesota, Kevilus, 62, says he often went "bushwhacking" as a child with his father. His childhood fascination with fire and camping led him to take up mule packing as an adult in Minnesota's remote North Woods.

Kevilus wanted to be able to do camping trips in the deep of winter but there was the conundrum of how to stay warm. "I have experience as a sheet metal worker and I had built some wood stoves for home use," he says. "So I decided to build a small, portable stove for use in a canvas wall tent." Soon Kevilus's friends wanted a stove, too.

In 1988, Kevilus and his wife founded Four Dog Stove Company in their hometown of St. Francis, Minnesota. Kevilus sells a variety of models that he has designed for tent-use, with some using titanium and others made from steel. All are portable and have a stovepipe, which nests inside the stove when it is being transported. Kevilus makes the stoves he sells by hand out of his St. Francis headquarters. In 2016 he sold nearly a thousand stoves, which caused him to stop taking orders until he could catch up with demand.

Kevilus applies his wood burning skills on camping trips as well as in his home that is also heated by a stove. "There is so much more to fire than just grabbing logs from the wood pile, putting them in the stove, and lighting it," says Kevilus. "I am still constantly improving my skills. I am always trying to see how efficient I can get with the fire building material I have on hand. It is a lifelong process of practicing the skills to get better at it."

air is heated and blown back out into the room. Meanwhile, combustion inside the insert is far more efficient than the old-fashioned fireplace because it uses modern technology to meet the EPA's strict emission guidelines.

NATURAL GAS

Another modern technology that gives the feel of a fireplace but without any of the work is natural gas heaters that can be inserted inside a fireplace or installed against any wall in a home as part of a faux fireplace. Natural gas heaters do not require a chimney because they vent laterally to the outdoors through an exterior wall. Most heaters also draw air through an outdoor vent as well. The "fire" is actually the combustion of natural gas fumes that shoot up above ceramic logs made to look like half-burned wood. The flame can be ignited and adjusted in size simply by turning a knob that controls the gas flow. The downside to these faux

fires is the cost, since natural gas is generally more expensive than wood for the same BTUs. And you don't get the same authenticity that comes from the crackling sound of a wood fire you started with your own hard work. The inner cave person in all of us probably finds staring at this artificial fire a little less satisfying.

PELLET STOVES

By far the easiest and most efficient way to burn wood, pellet stoves offer a nice compromise between the full automation of a natural gas insert and the somewhat labor-intensive wood stove. The stove requires electricity to run a mechanism that feeds wood pellets into a hopper for combustion. The heat generated by the fire is then blown out into the room through the top of the stove. The pellets, which are about the size of cat kibble, come in 40-pound bags that usually cost no more than $5 per bag because they are made from widely available wood waste. All that is required to run the stove is pushing the start button and pouring a few pounds of pellets into the top. Because of the various electronic parts, the stove must be periodically vacuumed to keep it working properly. Most pellet stoves are EPA certified and meet strict emission standards. Indoor air also stays clean with pellet stoves because there is no stray ash to blow out like with a wood stove. The downside to pellet stoves is the noise that the fan and pellet feeder make. And while the electricity required to run the stove is minimal it means that if the power goes out, the stove won't operate.

CHAPTER 6

Learning Your Way Around the Woodshed

All wood is not created equal—here's the lowdown on finding the best wood for your needs as well as how to stack it, store it, and split it

---

Until I was in my mid-thirties and I moved into that wood stove–heated mountain house in Flagstaff, I had never purchased a large amount of firewood. I had bought a bundle here and there from the campground store, but when it came to how to stock up with a supply of wood for the winter, I didn't have a clue. I was unprepared for the first autumn cold snap and bought several pre-packaged bundles of juniper from the grocery store. Then I bought a few more, and soon realized I would go broke if I didn't find a smarter way to supply my stove with wood.

It wasn't long before a man showed up at my front door with a pickup full of wood. He offered to sell me a cord, "at a very good price."

"How much wood is in a cord?" I asked.

This question was probably music to his ears, as I would later discover. Like most of the roving firewood salesmen in Flagstaff, this man was an independent operator who basically conducted business out of his truck. He and his associates cut wood from the national forest outside of town and then drove around looking for homes with a stovepipe or chimney where the occupants might be in need of more fuel. There were no government policies setting prices or controlling quality or quantity. It was up to the wood customer to be savvy in order to get the best deal. But that was not me.

He told me a cord "was a little more than a truckload" and that the wood he wanted to sell me was "mixed," as in some higher quality and some lesser quality. He named his price, adding that he had to be paid in cash. Was this a good deal like he insisted? I had nothing to compare it to, so I agreed to buy a cord. The man and his partner tossed all the wood from his truck into my backyard. I paid him and he drove away. He never came knocking on my door again, but I would sometimes see him in town, parked on the side of busy streets with a truckload full of wood.

The following year the next guy who came knocking on my door was more honest. When I bought a cord from him, I was amazed at how much bigger the pile of wood was that he threw from his truck, and how much better quality it was. That first cord was an expensive lesson. I vowed to never make the same naïve mistake again.

Over the course of a winter, I not only became knowledgeable about the amount of wood I was getting for my money, but also about the type of wood and what varieties would best serve my needs. I also learned how to stack it, split it, dry it, and when to buy it. Eventually, I became something of a wood connoisseur, appreciating rare varieties, as well as getting to know the best dealers, and stocking up on good deals before winter arrived. This knowledge and preparation made heating my home with wood a far more affordable and pleasurable experience.

Which is the Right Wood for You?

All wood is not created equal. Buying or harvesting the longest burning, highest quality, most expensive hardwood (even if price is no object), does not necessarily translate into the best wood for your needs. Quality and cost are factors but there are other considerations in much the same way you deliberate variables when deciding which car to buy.

Are you planning to use the wood for your only source of heat? In this case, you want wood that will burn as hot and as long as possible. Hardwoods with the highest BTU values (see chart on pages 118–20) should be the determining factor, but you will also likely be limited to the common tree species in your

geographic region. Apple, oak, and maple are among the hottest, longest burning woods and ideal for staying warm through the winter. However, where I live in Arizona, these are not native species, but juniper, a widely available endemic species to the western United States, is the wood of choice. Aspen is the No. 2 choice in my region. Once I got a bit smarter about buying firewood, I sought to purchase mixed cords of juniper and aspen. But I also kept an eye on my neighborhood in the event a tree trimmer was cutting a maple or apple tree in someone's yard. Discarded limbs from landscaping can make excellent—and sometimes free—firewood.

In New England, widely available maple is a wood stove staple. In the Midwest, it's more often oak and birch that are most plentiful and also provide optimum BTUs for surviving cold winters. In the Southeast, it's shagbark hickory that is prized for the way it produces maximum heat, along with a very pleasant aroma, but it is so dense that it can be hard to split. When considering what to use for winter heating, you not only want hardwoods that generate a high heat value, but that also produce long-burning coals to keep your fire going through the night.

However, if you are camping in mild weather and wildfire is not a threat, a few armloads of inexpensive pine could be just fine. You don't need to worry about the resin-heavy smoke when you are outdoors and oak or maple might burn too hot and too long for an evening gathering. On the other hand, if it's windy or you are burning indoors, a res-

inous wood like pine or spruce can be dangerous because it throws a lot of sparks and embers.

When the demand for BTUs is not an issue, lower density wood like willow or cottonwood will keep campers from overheating and these woods don't throw sparks. On the other hand, if you are planning to fish for trout on your camping trip and cook it, you will want to have a wood such as hickory, cherry, or apple that produces good coals and also adds a pleasant smoky flavor to food. And if you are renting a cozy mountain cabin, having wood that smells like the rustic atmosphere of the area—think juniper, cedar, or hickory—can add to the aesthetic of the place.

In addition to the BTU estimate, another good way to judge the potential heat output of a certain type of wood is how much it weighs when dried. "A wood's dry weight volume, or density, is important because denser or heavier wood contains more heat per volume," explains Michael Kuhns, an Extension Forestry Specialist and Professor at Utah State University. As an example, Kuhns points to the Osage orange tree, which weighs in at 4,728 pounds for a dried cord—and it has an exceptionally high heat output of 32.9 BTUs (plus it smells amazing). In contrast, a cord of dried white pine weighs 2,250 pounds and only puts out 15.9 BTUs. So even though white pine is usually far cheaper to buy, it is only worth half as much as orange or other dense hardwoods in terms of heating capacity. "Some firewood dealers sell mixed hardwood firewood. This may or may not be desirable, depending on the proportion

HARDEST
(MOST DENSE/
LONGEST BURNING)

Ironwood
Rock elm
Hickory
Oak
Sugar maple
Beech
Yellow birch
Ash
Red elm
Red maple
Tamarack
Douglas fir
White birch
Red alder
Hemlock
Poplar
Pine
Basswood
Spruce
Balsam

SOFTEST
(LESS DENSE/
LIGHTER WEIGHT)

of low-density hardwoods such as cottonwood that are included," cautions Kuhns. He also adds that ease of splitting should be considered because if splitting is an epic struggle, then super dense hardwoods may be too difficult to use.

How Much is a Cord?

Weight and density are key measures in choosing the best firewood for heating purposes. But when purchasing firewood, the standard measure for quantity is a cord, which is based on volume. A full (a.k.a. standard) cord when stacked measures four feet high, four feet wide and eight feet long. This translates into a volume of 128 cubic feet. However, part of this space is taken up by air in between all the pieces of wood. The actual solid wood content of a full cord is closer to 85 cubic feet.

The potentially confusing part about buying firewood is that dealers are not selling fuel logs in neatly stacked four-foot lengths. It is chopped up much smaller to be more useable and often haphazardly thrown in the back of a truck. I learned the hard way that a truck bed full of wood, even when it is piled up well above the tailgate, is not a full cord. A full-size pickup with a standard bed can hold about one half of a full cord—or 64 cubic feet—when loaded to the top of the bed. And since it is usually not stacked, there is air space, which means it may be even less than half of a full cord.

Another common measure used by firewood

Firewood Types and Attributes

SPECIES	HEAT PER CORD (MILLION BTUS)	EASE OF SPLIT- TING	SMOKE	SPARKS	COALS	FRA- GRANCE	OVERALL QUALITY
Alder	17.5	Easy	Medium	Moderate	Good	Slight	Fair
Apple	27	Medium	Low	Few	Good	Excellent	Excellent
Ash, Green	20	Easy	Low	Few	Good	Slight	Excellent
Ash, White	24.2	Medium	Low	Few	Good	Slight	Excellent
Aspen, Quaking	18.2	Easy	Medium	Few	Good	Slight	Fair
Basswood (Linden)	13.8	Easy	Medium	Few	Poor	Good	Fair
Beech	27.5	Difficult	Low	Few	Excellent	Good	Excellent
Birch	20.8	Medium	Medium	Few	Good	Slight	Fair
Boxelder	18.3	Difficult	Medium	Few	Poor	Slight	Fair
Buckeye, Horsechestnut	13.8	Medium	Low	Few	Poor	Slight	Fair
Catalpa	16.4	Difficult	Medium	Few	Good	Bad	Fair
Cherry	20.4	Easy	Low	Few	Excellent	Excellent	Good
Chestnut	18	Difficult	Medium	Moderate	Fair	Good	Good
Coffeetree, Kentucky	21.6	Medium	Low	Few	Good	Good	Good
Cottonwood	15.8	Easy	Medium	Few	Good	Slight	Fair
Dogwood	24.8	Difficult	Low	Few	Fair	Good	Good
Douglas fir	20.7	Easy	High	Few	Fair	Slight	Good
Elm, American	20	Difficult	Medium	Few	Excellent	Good	Fair
Elm, Siberian	20.9	Difficult	Medium	Few	Good	Fair	Fair

SPECIES	HEAT PER CORD (MILLION BTUS)	EASE OF SPLITTING	SMOKE	SPARKS	COALS	FRAGRANCE	OVERALL QUALITY
Fir, White	14.6	Easy	Medium	Few	Poor	Slight	Fair
Hackberry	21.2	Easy	Low	Few	Good	Slight	Good
Hemlock	19.3	Easy	Medium	Many	Poor	Good	Fair
Hickory	27.7	Difficult	Low	Few	Excellent	Excellent	Excellent
Honeylocust	26.7	Easy	Low	Few	Excellent	Slight	Excellent
Juniper, Rocky Mountain	21.8	Medium	Medium	Many	Poor	Excellent	Fair
Larch (Tamarack)	21.8	Easy-med	Medium	Many	fair	Slight	Fair
Locust, Black	27.9	Difficult	Low	Few	Excellent	Slight	Excellent
Maple, Other	25.5	Easy	Low	Few	Excellent	Good	Excellent
Maple, Silver	19	Medium	Low	Few	Excellent	Good	Fair
Mulberry	25.8	Easy	Medium	Many	Excellent	Good	Excellent
Oak, Bur	26.2	Easy	Low	Few	Excellent	Good	Excellent
Oak, Gambel	30.7	Easy	Low	Few	Excellent	Good	Good
Oak, Red	24.6	Medium	Low	Few	Excellent	Good	Excellent
Oak, White	29.1	Medium	Low	Few	Excellent	Good	Excellent
Osage orange	32.9	Easy	Low	Many	Excellent	Excellent	Excellent
Pine, Lodgepole	21.1	Easy	Medium	Many	Fair	Good	Fair
Pine, Ponderosa	16.2	Easy	Medium	Many	Fair	Good	Fair
Pine, White	15.9	Easy	Medium	Moderate	Poor	Good	Fair
Pinyon	27.1	Easy	Medium	Many	Fair	Excellent	Good

continues

SPECIES	HEAT PER CORD (MILLION BTUS)	EASE OF SPLIT-TING	SMOKE	SPARKS	COALS	FRA-GRANCE	OVERALL QUALITY
Poplar	13.7	Easy		Many	Fair	Bitter	Fair
Redcedar, Eastern	13	Easy	Low	Many	Poor	Slight	Fair
Redcedar, Western	18.2	Medium	Medium	Many	Poor	Excellent	Fair
Spruce	15.5	Easy	Medium	Many	Poor	Slight	Fair
Spruce, Engelmann	15	Easy	High	Few	Poor	Slight	Fair
Sycamore	19.5	Difficult	Medium	Few	Good	Slight	Good
Walnut, Black	22.2	Easy	Low	Few	Good	Good	Excellent
Willow	17.6	Easy	Low	Few	Poor	Slight	Poor

Source: "Heating With Wood: Species Characteristics and Volumes," by Michael Kuhns, Extension Forestry Specialist, and Tom Schmidt, Forester, Utah State University Extension Forestry. For more information, including characteristics of tree species, and other firewood advice and science, go to Utah State University's Extension Forestry website: forestry.usu.edu.

dealers is face cord or short cord. This is a stack of wood that is four feet high, eight feet long, and as deep as the pieces are long—usually about 16 inches. Since the pieces are cut to fit inside a wood stove or fireplace, this brings the volume down considerably when compared to the full cord measure of four feet wide. A face cord is about 1/3 the volume of a full/ standard cord, and there are approximately 220–240 pieces of wood in a face cord. If a dealer drops a full cord in your yard, the pile should be about eight feet in diameter and eight feet tall.

There are some regional variations in measure used by firewood dealers. For example, in Vermont, a face cord is sometimes called a "rick." Elsewhere in New England it is also called a "run." In New York City where people live in high-rise apartments with nowhere to store wood outside, there is the "city cord." This is wood sold in bundles—25 bundles is half a city cord.

Firewood dealers also commonly sell "mixed" hardwood cords. Buying a mixed cord is almost always cheaper than buying a full cord of premium hardwood like maple or oak. The mixed cord will have less dense woods like aspen or birch along with the maple and oak. The lower density woods can serve a purpose in terms of being easier to ignite in

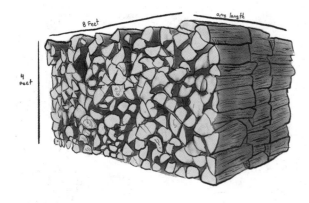

8 Feet

any length

4 Feet

A face cord, which is less than a full cord

order to get the fire going and also burning during milder times of year. But make sure you are not being short-changed in BTUs with a mixed cord that has too much of the lighter weight fuel.

So, how much wood do you actually need? If you are burning one to two fires per week, one face cord should last the season. If you are heating your home with wood, you can expect to burn two to four full cords per year, depending on the climate and size of your house.

Buy Local

The idea of burning an exotic wood imported from a far-away forest may sound appealing if your local firewood options seem boring. But doing this could be disastrous for your neighborhood or the

area where you are camping. An insect infestation impacting one species of tree in a certain region is easily spread if the infected wood is transported to another region. Beetles or other insects from just a single piece of infested wood can migrate from a backyard woodpile and sicken trees throughout a neighborhood or recreation area.

"Try to use firewood that is growing as near as possible to where you are using it," advises Michael Kuhns. "Even if the only wood available in your area is pine that produces a lot of smoke and sparks, that is safer for the environment than importing wood from 100 miles away." Kuhns adds that insect infestation is especially a problem in moist, humid climates. In these places he suggests obtaining firewood within a 50-mile radius of where you plan to use it. In drier areas such as the intermountain West, he advises venturing no more than a hundred miles to get firewood.

States in the mid-Atlantic, New England, and Midwest regions have implemented restrictions on firewood harvesting and sales in an effort to control

the spread of harmful insects and exotic species. In certain hard-hit areas, infestations have destroyed forests as well as trees in neighborhood parks. New York, Pennsylvania, Ohio, Michigan, and Wisconsin prohibit the sale of green firewood outside a 50-mile radius from where it was harvested. However, the New York Department of Environmental Conservation allows firewood that is first dried in a kiln—killing the insects—to be moved more than 50 miles from its source as long it is first certified and labeled as "New York Approved Treated Firewood/Pest Free."

Most public lands, with the exception of national and state parks, allow harvesting of firewood with what the U.S. Forest Service calls a "Personal Use Permit." Rules and conditions for cutting wood vary by region. Cutting wood without a permit from federal and state public lands is illegal. To find out more about personal use permits, go to the U.S. Forest Service website. You should also investigate the regulations posted by the state department of environmental quality in your area before harvesting wood.

For Every Wood There is a Season

Equipped with moisture and nutrient sucking roots that extend far into the ground, trees are designed to hold water like a sponge. When first cut, a tree's weight is comprised of at least 50–60 percent water.

Trying to burn wood that is so full of moisture not only makes it hard to ignite but it also produces an unpleasant amount of smoke and creosote. No matter how high the projected BTU rate is of a certain wood, it must first be dried (a.k.a. seasoned) to function efficiently. However, this does not mean that all the moisture content must be eliminated. Wood that has reached a moisture content of 20 percent or less is considered to be well seasoned. But this does not happen overnight. Drying wood requires sufficient time as well as exposure to air and heat from sunlight.

HOW LONG?
"You will get the best firewood deals if you are organized and plan ahead," advises Kuhns. "You can buy green wood much cheaper than dried wood but you will need to buy it nine months to a year in advance and dry it in your yard."

A general rule for drying time is that the wood should be cut in spring before the next winter. It is important that the wood is properly stacked so it has maximum air and heat exposure during the summer months. If wood is cut even a year or two years earlier but sits in a giant pile instead of being stacked, it will retain too much moisture for efficient use. On the flip side, seasoned wood never goes out of season. If stacked properly it can sit for years off the ground and still be usable. Keep in mind that it takes longer for hardwoods to dry than softwoods because of the greater density.

And wood split into smaller pieces will dry more quickly because of the increased surface area that is exposed to air and heat.

EXPOSURE

In order for air and sun to dry the wood, it must be exposed to the elements. As soon as a tree is felled, the logs should be cut down to pieces that are the right size to fit into a wood stove or fireplace (at least 3 inches shorter than the length of the firebox). And then those logs should be split in half to open them up to air and sun. Next, the wood should be stacked in a long, narrow row so that split logs are exposed to the elements on both ends as well as air circulation coming down from the top (see Stacking and Storing Tips on pages 126–28). If it's a full cord, the stack can be about four feet tall and positioned in a location away from the house that receives plenty of sun and wind. In moist climates, a tarp may be placed over the top of the stack but not the sides. In arid regions, many wood heating veterans don't cover their stacked wood at all. This theory holds that precipitation only penetrates the top layer of the stack and the extra air circulation the stack gets from not being covered compensates for the drenching from summer rain.

IS IT DRY ENOUGH?

Most firewood dealers market their product as already seasoned. But how do you know the cord you are paying several hundred dollars for has really been properly dried? Here are some telltale signs indicating wood is seasoned.

Weight: Pick up a piece of "dried" wood and compare it to a similar size/type of green wood. The dried wood should feel substantially lighter.

Cracks: Look at the ends of cut pieces. Well-aged wood will have what is called "checks," which are fine cracks radiating out from the center of the round log. These cracks are also excellent axe entry points for splitting.

Loose Bark: As the wood dries, the protective layer of bark loosens and eventually falls off. Bark that sloughs off a log when you pick it up is a good sign.

Appearance: The seasoning process sucks the color and aroma from wood. The more faded, the better.

Rings Hollow: If you hit two pieces of green wood together it will make a solid thud sound. However, two pieces of seasoned wood should make a ringing sound.

WHAT ABOUT DOWNED WOOD?

Picking up dead, downed wood where it is legal and available to harvest near a public campground or residential area can be a fine source of fuel. However, unlike stacked and split firewood, the wood on the ground can become too rotten to be good fuel. If a log has lost all its bark and falls apart when you pick it up, then the decomposition process is far enough along that it will not burn well.

Stacking and Storing

There are two stages of storing wood after it has been cut. The first is stacking for the purposes of drying. The second is long term winter storage after the wood has been fully seasoned.

HOW TO STACK

Storing fuel logs for seasoning can be as simple as stacking them in a single row along a fence or it can be turned into an artistic backyard sculpture—a popular novelty in cold European places like the Swiss Alps. No matter how you choose to stack it, the three priorities should be keeping the wood off the ground, exposing cut sides to air and sun, and stabilizing the stack to prevent it from falling over.

The foundation of the stack can be large uncut logs that you sacrifice to ground moisture or you can use standard wooden pallets. Some type of bookend—such as a metal pole or wooden two-by-fours—can be driven into the ground on the ends of the stack structure to hold it in place. Logs on the ends of the stack can also be bundled together in groups of three with twine to reduce shifting in the pile. Avoid using trees for bookends because insects from the trees can move to the woodpile or vice versa. Stacking wood against a house is a poor choice under any circumstance due to the propensity for termites to migrate to the woodpile and then into the structure.

Once you have your foundation on level ground, begin stacking pieces in parallel fashion and as

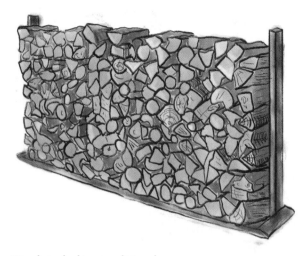

Wood stacked in a traditional manner

Wood stacked using cribbing method

close to each other as possible. The stacked wood should be in single rows that are no more than about four feet high. Throw knobby or bent pieces aside so they don't compromise the stable symmetry of the stack. These outliers can be placed on top. While positioning all the stacked pieces parallel in the same direction works just fine, there are some wood-heating aficionados who are proponents of "cribbing." This method creates Lincoln Log–type towers where each ascending row of wood is placed perpendicular to the row beneath it. The philosophy of cribbing is that the cross-hatch style affords more air and sun exposure to the wood.

STORAGE

For people who are heating with wood in a cold climate or who just have a lot of firewood there needs to be a sheltered storage area for the wood after it has been adequately dried. Because of the potential termite and insect problems, storing wood on a front or back porch is a bad idea even though it may be convenient when feeding the wood stove on cold nights. A wood shed is ideal and can be simple to build. The goal with a shed is to shelter the wood but still allow air circulation so the wood stays dry and free of mold.

The shed should be located a convenient distance from the house and in a place that receives good air circulation. It should also be big enough to hold a winter's worth of wood and accommodate storage of different types of fuel in designated compartments—for example, a spot for kindling, an area for softwoods, and the bulk of storage for hardwoods. The fuel in

the shed should be organized so that you can quickly dash in, get what you need without causing an avalanche of stacked wood, and then run back into the house. In order to prevent the possibility of mold contaminating indoor air quality, only keep one to two days' worth of wood inside your house.

A simple, functional shed design is one that has three sides and a pitched roof. The base can be either poured concrete or wooden pallets surrounded by poured concrete footings for posts. The top of the walls can be left slightly open to increase air circulation. There should also be a dedicated space in the front of the shed for splitting wood. A functional and well-organized shed makes storing wood outside the house easy.

The Art of Splitting

Splitting wood is not just the domain of burly lumberjacks in flannel shirts. Anyone can learn to do it with practice. I can personally attest that taking an axe to smaller logs or wood pieces is not so much about strength as it is precision, skill, and confidence. Plus, it helps to have the right tools.

TOOLS

- Long-handled (28–34 inches) axe that is sharpened frequently
- Maul for splitting larger logs
- Splitting wedge to use with maul on difficult pieces
- Hatchet for splitting apart small pieces of kindling

TECHNIQUE

Splitting wood on the ground is dangerous because of the potential to chop off your toes. Use a large flat-surfaced log or stump for a chopping block. Make the chopping block appropriate for your height. You want to be able to split the wood with a relatively straight back.

Place the wood to be split standing vertically on the chopping block. Aim for the center of the wood and cracks or checks that are an easy entry point for the blade. If you find that your aim is not great and the wood goes flying off the block without being split, use the tire trick. Place an old tire on top of the chopping block and put several pieces of wood standing inside the tire. As you swing at the pieces, the tire will hold the wood upright and on the block.

There are two different ways to swing an axe. You can hold it up over your head and come down directly on the wood with an explosion of force that is powered by the upper back. Or you can build momentum by beginning your swing from the side, winding up, and then coming down from overhead. This torque of the torso is similar to throwing a baseball.

For large logs, a heavier maul is more effective than an axe. You may also use a splitting wedge and drive the wedge into a crack with the maul. Then swing the maul down on the wedge to break the log apart.

Be sure to collect wood debris of all sizes cast off during splitting for tinder and kindling.

FIRE KEEPER: MICHAEL KUHNS

As the head of the Wildland Resources Department at Utah State University, forest ecologist and professor Mike Kuhns has devoted his professional career to helping people better understand and appreciate trees. In addition to publishing extensively on ways to improve forest conditions in the wildland-urban interface, Kuhns has compiled a comprehensive guide to tree species in the United States for Utah State University's Forest Extension program. And this includes how to best use trees for wood heating.

Kuhns' personal philosophy on wood heating is that fuel should be harvested and prepared with the lowest carbon footprint possible. "Driving 100 miles to get firewood or paying a dealer for wood that has been dried in a kiln makes absolutely no sense to me," he says. "You should expend as few BTUs as possible to get the BTUs out of your firewood. You should live with what you've got where you are."

Early in his career Kuhns was working for the Medicine Bow National Forest in the tiny town of Encampment, Wyoming. And his only source of heat during the brutal winters was a wood stove. He burned aspen nonstop because that was what was locally available.

"There's a saying that everything looks like firewood to a person who burns wood for heat," he jokes. "That was how it was for me in Wyoming. If I saw a big log laying on the side of the road when I was driving I would pull over, pick it up, and put it in the back of my truck." Kuhns says he even burned old furniture.

And he was not alone. When the Forest Service surveyed visitors in the Medicine Bow National Forest about their favorite recreational activities, collecting firewood was at the top of the list for most people. "We didn't even have it on the form as an activity to choose from," says Kuhns. "People wrote it in."

CHAPTER 7

Cooking with Fire

Learn how to harness flames for delicious outdoor meals with these tasty recipes and centuries-old cooking methods

---

Consider this the next time the wafting aroma of meat cooking over open flame makes your stomach growl: Humans have been cooking animals and vegetables for nearly two million years. And only in the last century has cooking taken place on gas or electric ranges. Fire is not only synonymous with the act of cooking but also with eating. Your gut-brain connection remembers this evolutionary fact even if you mistakenly believe the microwave is what sustains you.

However, fire historian Stephen Pyne points out that cooking alone was just one part of the vital relationship fire played in feeding humans. The earliest human ancestors likely scavenged their first cooked meal from animals killed by wildfire: "Fire cooking followed fire hunting, fire foraging, fire-based farming, and herding. Fire helped ready meat, grain, or tubers for eating, improving the taste, leaching away toxins, killing parasites. Fire—its heat, its smoke—then helped preserve what was not instantly eaten."

Whether your fire is in a campground, the backyard, or an indoor fireplace, utilizing the heat energy for cooking not only offers a primal satisfaction but is also a great way to add extra value to the wood you are burning. If you are already sitting entranced in front of the fire for an hour or more, why not also stare at your dinner cooking? Plus, it is a great way to add flavor, as many gourmet chefs who utilize pan searing and fire cooking methods can attest.

Many of the best meals I have ever eaten were enjoyed while sitting around a campfire. It's true that many things taste better simply because you are outdoors and exhausted after a long day of hiking or other activities. But there is something truly pleasurable in and of itself about food cooked over a fire in a primitive setting. As I watch the sunset and chat with friends and family around the fire, I feel like I am able to relish every bite.

The same low-tech techniques used for hundreds of thousands of years to cook over a fire continue to stand the test of time. Basically, cooking with fire either happens over open flame or on a bed of coals or ashes. The cooking methods using fire are essentially the same as those employed when preparing food with a gas or an electric stove. You are either baking, roasting, grilling, or boiling. But unlike with a modern appliance, you may also be toasting a marshmallow and telling a story while sitting around the flames cooking your dinner.

COOKING WITH COALS

A bed of glowing coals is the original stovetop for a variety cooking needs, from grilling a steak to roasting a potato to baking bread. Large fireplaces used by America's early European settlers had different compartments for containing coals, often with a shelf in the back for baking and an area off to the side where coals could be corralled without limiting the fire's role in heating the home.

The "keyhole" design is the tried-and-true method for burning wood that will serve both warming and cooking purposes. A keyhole fire has a circular fire pit where fuel logs are burned but there is also a smaller rectangular area extending from the circle perimeter (hence, the keyhole shape) where coals from the main fire are shoveled and nurtured into a flat, sizzling bed.

The ideal fuel for producing cooking coals is hardwood such as oak, hickory maple, pecan, and

cherry. This higher-density wood not only burns longer but retains a hotter, consistent heat that is ideal for stable bed of cooking coals. Plus, the aroma from these woods adds a pleasing smoky taste to foods. Softwoods such as pine not only have shorter burn times but the resin can infuse foods with an unappealing flavor. Expect to wait about 45 minutes to an hour after the fire starts to get the first coals sufficient for cooking.

Use a rake or shovel to move the red lumps of char into the rectangle of the keyhole. Make the coal bed 10–12 inches away from the flaming part of the fire so that it is not impacted by the bursts of heat coming from the combusting fuel logs. Keep in mind that even though the coals are not flaming they are still very much on fire. Add new coals to the bed every 20–30 minutes to keep up consistent heat. This means that it could be an hour and a half from the time you start the fire to when the coal bed is big enough for cooking.

As for the size of the coal bed, the cooking surface does not need to be much larger than the circumference of the pot or griddle you are using—just as with a stovetop burner. To keep the heat energy even, use metal tongs to redistribute coals and place newer, hotter ones among those that have been sitting for a while. If you plan to cook for a large group or prepare multiple courses, you can also keep the coal bed burning longer and more steadily if you make it two or three layers deep.

But how do you know when the coals are hot enough for cooking or grilling? Rather than risk burning your food, use the toaster test. Toasted bread turns a golden brown at about 310°F. Gauge the coal bed heat by placing a slice of bread directly on the coals or in a pan atop the coals. If the bread toasts perfectly within a minute or so, then you not only know the approximate temperature but you also have the makings for breakfast or a grilled cheese sandwich. Another heat gauge method is to simply hold your hand a few inches above the coals. Keep your palm facing up because the top of the hand is more sensitive to temperature fluctuations. Time how long you can hold it there before it gets uncomfortably hot.

1–2 seconds translates to a coal bed that is
    450–500°F
4–5 seconds, 350–400°F
7–9 seconds, 250–300°F

Grilling

Placing a sturdy metal grill—the kind used on portable barbecue pits—about 8–10 inches above the coal bed is an easy way to cook or grill foods. If you want to prepare multiple items at once, you can grill meats on a hotter area of the coals and cook side dishes in a pot or foil atop a cooler area. Keep in mind that heat from coals can be more intense than a regular stove so you will need to flip the meat and stir the pot often to keep things from burning. Many established campgrounds have fire rings with grates already installed. However, if you

are cooking in a backyard fire pit or somewhere off the grid, you can fashion grill supports from two large, green fuel logs or sturdy, flat rocks. If you are camping, use green wood (to be burned later) rather than damaging the environment by leaving pristine rocks with black fire scars. Ready-made adjustable grills are also handy (see Resources on page 154).

When it comes to grilling meat, opt for sturdy, thick cuts with little fat in order to minimize dripping onto the coals and causing flames. A sturdy,

level grate is also a great stovetop for holding multiple pots and pans.

However, if your paleo instincts are burning strong, you can also opt to cook a steak directly on the coals. This is best when you have high quality wood with a flavorful aroma, such as hickory, maple, or cherry. Use a bone-in cut that is about 1½ inches thick. Before placing the steak on the coals, blow lightly on the bed to remove ash. Then cook for approximately 10 minutes on each side. This method is a favorite of chef Tim Byres, an evangelist

Espresso-Chili Dry Rub

Makes about ¾ cup

Coat a fine cut of meat in spices, throw it on the coals and kick back. This caveman cooking method is as simple as it is delicious.

INGREDIENTS

1½ teaspoons finely ground espresso coffee

¼ cup ancho chile powder

¼ cup dark brown sugar, tightly packed

1 teaspoon dry mustard

1 teaspoon ground coriander

2 tablespoons kosher salt

1 tablespoon ground cumin

INSTRUCTIONS

1. In a small bowl, mix the ingredients thoroughly, massaging the mixture with your fingers to break down the dark brown sugar into fine crystals.

2. Liberally sprinkle a thin layer of the rub onto the steak, then pat in with your fingers so it adheres to the surface.

Skillet

From flat breads to pizza to a breakfast egg scramble, the most delicious camp meals can come from a simple skillet placed on or above a bed of coals. As for fire-worthy cookware, whether it is a skillet or a pot, cast iron is the most durable and radiates heat to the food most effectively without burning it. Cast iron has been used to cook with fire for hundreds of years and a high quality iron skillet is well worth the investment. One of the best cast iron manufacturers in the United States is Lodge. The company was established in 1896 in South Pittsburg, Tennessee, by Joseph Lodge and has been turning out quality cookware ever since (see Resources on page 154).

For the best results oil and heat the skillet before cooking. If possible, place the skillet on a grate above the coals to control heat when cooking dishes made with eggs, cheese, and sauces. However, bacon, potatoes, and other vegetables are often tastiest when grilled wok-style in a skillet sitting directly on top of sizzling coals.

Fun with Foil

Foil and campfires go together like peas and carrots. Just about any kind of meat and vegetable can be rolled up into foil and tossed onto a bed of coals for convenient cooking. Adding seasoning can make the ingredients tastier thanks to the way spices infuse the food during the foil cooking process. And

for cooking with fire in his five-star, Texas-based Smoke restaurants. Byres recommends patting the steaks with a dry rub before cooking to protect the meat's surface from the char and infuse it with extra flavor.

because of the simplicity of preparation and cooking, along with the ability to individualize what you put in the foil, this is a great way for kids to have fun making their own dinner on camping trips.

Use 12-by-18 inch sheets of heavy-duty aluminum foil and wrap loosely around your ingredients, tightly rolling up all edges to create a pouch. The pouch should be roomy enough for the juices to escape and become part of the cooking/seasoning process. Place on a coal bed that is at least 3 inches deep and about medium heat. For the recipes described here, the coals are ready when you can comfortably hold a hand 6 inches above the coals for 3–4 seconds. For cooking dense, uncut vegetables like potatoes or squash put a layer of coals on top of the pouch as well. Then, be patient. This is not a microwave.

Foil cooking is a long-time camping tradition. Here are a few time-tested foil favorites:

Stew: Cut 4 ounces of beef or lamb into cubes. Thinly slice a potato, carrot, summer squash, small onion, and garlic. Arrange the slices on the foil. Sprinkle with salt. Add several tablespoons of water and fold up the foil. Cook on the coals for about 20–30 minutes.

Bacon Hamburger: Shape 4 ounces of hamburger into a patty. Cut a medium-sized potato and carrot into thin strips. Peel and slice a small onion. Lay two bacon strips across the patty with the veggies on top. After arranging all the ingredients on a square of foil, sprinkle lightly with salt. Close the foil, lay the package on the coals, and cook for about 15–20 minutes.

Fish: Wrap fresh fish in foil along with some finely chopped onion and lemon. Slice a yellow pepper and place strips on top of the fish. Bake on the coals about 5 minutes per side for a smaller fish, or 10 minutes for a large catch.

Baked Potato: Pierce the skin of the potato in several places, then wrap in foil. Bury it deep in the coals for about 30–40 minutes.

Fruit: Cut the core out of a raw apple and replace it with a pat of butter, a few raisins, some cinnamon, and a teaspoon of brown sugar. Wrap in foil and bake on the coals for 30 minutes.

Don't be fooled by the simplicity of foil cooking. It is entirely possible to be a foil gourmet. Here is just one idea of countless options:

From the Ashes

Long before there were ovens or even utensils, humans cooked many items simply by burying them in a bed of hot ashes. The food items that are ideal for this kind of cooking are generally high density with a moderate amount of internal moisture. They also have a fairly tough outer layer that protects the insides and can be peeled away. Roasting food in ashes is a slow process and should be done when the coals have produced a pile of spent ashes that are still hot. Use a shovel to push the ashes into a thick bed that is next to still-burning coals.

Foods that are prime for ash roasting are whole potatoes, sweet potatoes, winter squash, onion, garlic, sweet corn with the husk on, plantains, peanuts, and chestnuts.

The steady heat produced by a bed of ashes also makes a fine oven for baking bread. If you've ever watched western movies you have probably seen this in action with cowboys sitting around a campfire making flat breads called "ashcakes" or "hard tack." Once the bread is done, the ashes can be brushed off, the cake buttered and enjoyed fresh and hot from the "oven."

Even though the historic ingredients used for ash cakes sounds simple enough—flour, lard, baking powder, salt—anyone who has tried to make them probably ended up with inedible globs more often than not. The easiest way to get around this is to take advantage of the convenience of Bisquick or instant pancake mix. Either works fine.

Southwest Campfire Scramble

Serves 2

A hot, cast-iron skillet and basic ingredients are all that is needed to start your morning with a hearty eggs-and-sausage breakfast.

INGREDIENTS

4 oz. Italian sausage

2 small jalapeno peppers, diced

1 Anaheim pepper, diced

½ sweet onion

5 eggs

Kosher salt and freshly ground black pepper

6 oz. Monterey jack cheese, crumbled

½ cup cherry tomatoes, sliced in half, or 4 Campari tomatoes, cut into quarters

Salsa

Tortilla chips

INSTRUCTIONS

1. Heat skillet on coals. Place skillet on grate above coals and add sausage, peppers, and onions. Cook over medium-high heat until browned, breaking meat into small chunks. Remove and set aside.

2. Meanwhile, in a large bowl, whisk eggs with a little water in bowl, and season with salt and pepper.

3. Pour eggs into 8-inch heated cast iron pan, stirring occasionally until they're halfway set. Stir in sausage/pepper/onion mixture, along with crumbled cheese and tomatoes. Cook over coals, stirring gently a couple more times, until cheese melts and eggs are almost set. Garnish with salsa and crumbled tortilla chips.

Pepperoni Pizza Logs

Serves 4

Prepare your pizza days or weeks ahead of time and stick it in the fridge. Once in camp, place the foil-covered rolled dough on hot coals, and, presto! Dinner is ready in minutes.

INGREDIENTS

2 sheets heavy-duty foil; double them
    and coat with a light layer of olive oil

14 oz. refrigerated pizza dough

12 oz. pizza sauce

12 oz. shredded mozzarella cheese

8 oz. thinly sliced pepperoni

Dash all-purpose flour

Dried Italian seasoning

Personalize: Add your own favorite fixings

PREPARATION AT HOME

1. On a lightly floured surface, roll pizza dough into a quarter-inch-thick square. Spread pizza sauce evenly over dough to within ½ inch of edge. Top with cheese and pepperoni (adjust amount to taste). Sprinkle with Italian seasoning. Starting at one end, roll pizza into a log. Transfer to prepared foil and wrap tightly, sealing all edges. Refrigerate for up to three days or freeze for up to two months.

AROUND THE CAMPFIRE

2. Thaw pizza log if frozen. Prepare campfire coals. Place wrapped pizza log on hot coals and cook for 15 minutes. Turn log over and cook for 10–15 minutes more or until crust is golden brown and log is heated through. Remove from coals, cut foil open, and let rest for 5 minutes before slicing.

Instant Ashcake Recipe

Serves 4–6

Old fashioned cowboy cooking is much easier and tastier if you cheat just a little by using instant baking mix. Garnish with berries to add a punch of flavor.

INGREDIENTS

½ cup water

2 cups instant pancake mix or Bisquick

INSTRUCTIONS

1. Add water to instant pancake mix or Bisquick until you have a handful of squishy yet slightly firm dough.

2. Pat the dough into a quarter-inch-thick pancake. (In order to keep the pancake from sticking to your hands, sprinkle hands with a little dry mix before patting.)

3. Place the pancake into a bed of ash-covered coals and watch the dough start to fluff up.

4. Cook on one side for about 1–2 minutes. When the dough becomes slightly rigid and the edges start to turn brown, flip the cake with a spatula. Cook on the other side for about 30–60 seconds.

5. Remove the cake from coals and let it cool for a few seconds. Blow away any lingering ash.

6. Slather with butter, jam, honey, or maple syrup. Add fresh berries if you have them.

Peach Streusel Coffee Cake

Serves 4–6

Learn to bake in a Dutch oven

with this simple yet sumptuous recipe that is a decadent camp treat for breakfast or dessert.

CAKE INGREDIENTS

1¾ cup flour

½ cup brown sugar

⅓ cup powdered milk

1 tablespoon powdered egg

2 teaspoons baking powder

½ teaspoon salt

4 tablespoons margarine, melted

1 teaspoon vanilla

½ cup water

1 (10-oz.) can of peaches, drained

TOPPING INGREDIENTS

5 tablespoons brown sugar

⅓ cup oatmeal

4 tablespoons margarine, melted

1 teaspoon cinnamon

INSTRUCTIONS

1. Grease and flour Dutch oven.

2. In a separate container, mix dry ingredients; add liquids and stir.

3. Pour mixture into pan. Place peach slices on top of batter, arranging in concentric circles going from the pan's outer edge toward the pan's center. Mix topping ingredients in a separate bowl and pour on top of peaches.

4. Bake over coals 15–25 minutes, using a twiggy fire on the lid.

Backcountry Baking

One of the most valued possessions in the Colonial American home during the eighteenth century was the family's Dutch oven. Named for the country that developed a casting method producing smooth metal surfaces, this thick-walled pot with a tight-fitting lid was the primary vessel for baking, frying, boiling, and roasting. Silversmith and American Revolutionary War hero Paul Revere is credited with improving the Dutch oven. He added legs so the pot could sit above fireplace coals and a flat lid with a rim that made it easier to place coals on top during cooking. The heavy cast iron pots were packed on the Lewis and Clark expedition as well as on wagon trains carrying pioneers heading west who wanted their campfire cuisine to be a step above ashcakes.

Like a regular oven, the Dutch oven bakes by surrounding the food with consistent heat. The design has changed little since the eighteenth century. Breads and cakes rise inside the oven because coals are placed on top as well as underneath and on all sides of the round pot. A small fire with twigs (a.k.a. twiggy fire) can be built on the lid to improve baking performance. Operation of a Dutch oven requires a large supply of coals to keep the heat steady but nothing beats having fresh baked bread or hot cobbler on a camping trip. With a Dutch oven, even a backcountry birthday cake is possible.

Other Dutch oven tips:

- Generously grease the pot, including the inside of the lid, so dough does not stick to the surface.
- Only fill the pot halfway to leave room for dough to rise.
- Rotate the pot every few minutes to compensate for the uneven heating nature of cooking on coals.

FOOD OVER FLAME

Roasting Spit

Perhaps nothing is more primal than roasting an animal over a fire, caveman style. When you have an entire chicken or other animal to cook, rotating it over open flame or hot coals is efficient and will produce juicier meat than if it was sliced into pieces and grilled.

A roasting spit can be built over a fire pit with wood gathered from the surrounding area. For the spit and skewers that hold the animal, choose green wood that is fairly straight and, ideally, from a tree species that has little aroma or a pleasant smell, such as cherry. As the animal cooks, the smell of the wood will inevitably infuse the taste of the meat. The spit should be at least three times as long as the animal, while the skewers should be a few inches wider than the broadest part of the animal. In addition to the spit and skewer materials, find two sturdy branches with y-shaped crooks that can be stuck into the ground and hold the spit.

Use a knife to shave the spit stick smooth, cutting off any twigs and carving a pointed end. Also carve the smaller skewers sharp and smooth. Run

Bringing Things to a Boil

When it comes to quickly bringing water to a boil or maintaining a healthy simmer for cooking, the easiest way to do this is over open flame instead of coals. The flaming part of a fire is more than 1,000°F and water boils at 212°F, so the morning coffee will come much quicker from a pot placed over flames.

Like the Dutch oven, pots designed for use in fire utilize technologies from the eighteenth century or earlier. Most pots are made of cast iron or lighter weight anodized aluminum. They should always have a bail handle to allow safe removal from the flame using a stick or other utensil. Some pots and kettles are double walled or have nesting cups to allow for a double boiler and/or reduce the likelihood of food burning in the fire (see Resources on page 154 for suppliers).

Rather than putting the pot into a large fire, it is better for cooking purposes to build a smaller fire off to the side where the heat can be more easily controlled and the flaming area is roughly the size of the bottom of the pot. The pot should be positioned about 6–12 inches above the fire. Portable grills with adjustable heights are handy for this purpose (see Resources.) Or, if you can find three similar sized flat rocks, position them around the small fire, and balance the pot on the corners of the rocks. However, be mindful in campgrounds or other public lands of not scorching rocks and leaving an unsightly mess for the visitors who come

the spit stick through the animal, from head to tail, along the underside of the backbone. Run the skewers perpendicular to the spit, through the animal's wings or legs, to stabilize it when rotating the spit.

Bury the spit support sticks in the ground so that the animal can rest about one foot above the coals. Use rocks to stabilize the supports so you don't have a tragic collapse. Balance the spit over the fire and rotate the animal every three to four minutes. Place a drip pan underneath to catch grease and prevent it from igniting flames and scorching the meat. The drippings can also be used for a sauce or marinade simply by frequently stirring, and adding a bit of wine and seasonings.

Be patient. A slow roast to produce the juiciest meat can take anywhere from 75 minutes to two hours, depending on the size of the animal.

after you. A method used by the Boy Scouts is to create a pot rest with three large L-shaped metal tent stakes that are stuck in the ground in the middle of the fire. Another technique is to create a lever from which the pot is suspended over the fire, but this requires a pot with a tight fitting lid and sturdy handle.

For the lever method, take a sturdy, fairly straight branch and position it over a large log so that one end of the stick is above the middle of the fire. Choose green wood, which is less likely to burn. Anchor the stick in place on top of the log with a few heavy rocks, causing the stick to angle nicely above the fire. Place the pot handle on the far end of the stick. Carving a notch in the stick to hold the handle will make it more secure. Experiment with an empty pot to make sure the rig is stable before you risk losing your dinner in the fire if the lever collapses.

Dinner On A Stick

Putting food on a stick and dangling it over flames was likely the first cooking method and it has never gone out of style. Whether you are toasting a marshmallow or roasting a kebob, this simple technology remains one of the most pleasurable ways to enjoy a fire as well as food. It is fast and fun, especially for kids. The reason chunks of food on a stick cook faster than an animal on a spit has to do with surface area. Just as the increased surface area of a handful of pine needles makes the tinder far more flammable than a large fuel log, cutting food—especially meat—into smaller pieces gives it more surface area to be heated in cooking.

Campfire kebobs can be made with any foods that are stable enough to withstand piercing and rotating over flames without falling apart. Items should be cut into uniform squares or slices and skewered through the middle. As for the skewer, use metal or a sturdy green stick that has been shaved smooth with a knife. Meats—chicken, beef, or lamb—should be marinated for extra flavor. Other excellent kebob items include mushrooms, pineapple, cherry tomatoes, small red potatoes, green or red peppers, onions, slices of corn, and zucchini. Slowly rotate the kebobs over the fire or place them on a grill above the fire and rotate every few minutes.

Spicy Beef, Red Pepper, and Mushroom Skewers

Serves 6–8

Cooking on a stick is as basic as it gets. But this recipe for a spicy beef marinade and mix of vegetables turns paleo practices into campfire gourmet.

INGREDIENTS

1 tablespoon ground black pepper

1 tablespoon ground coriander

1 tablespoon ground cumin

4 cloves garlic

1 tablespoon chopped Serrano pepper

2 teaspoons kosher salt

½ tablespoon balsamic vinegar

1 beef roast (about 2 pounds), cut into 1-inch chunks

1 large red bell pepper, cut in 1-inch bits

12 oz. button mushrooms, halved or quartered to similar size as pepper and beef

½ cup cilantro leaves and stems, roughly chopped

INSTRUCTIONS

1. Mix spices and Serrano in a bowl. Add garlic, red pepper, and salt, and pound to a rough paste. Work in the oil and vinegar, mashing it all well together. Massage the mixture into the beef chunks and cover, keeping it at about room temperature. (If camping, do this preparation ahead and keep refrigerated beforehand.)

2. Thread meat and vegetables alternately on skewers. Grill over fire or coals, rotating frequently and allowing food to turn brown on all sides.

3. Slip off skewers and toss with chopped cilantro before serving.

The Perfect S'more

Since marshmallows have such high sugar content, they are a prime confection for fire roasting. If toasted well, the outer marshmallow will be caramelized in the same way a gourmet chef tops off flan with a crispy sugar glaze. Sure, there are always people who insist they prefer their marshmallows charred black. But maybe the truth is they just don't know the proper toasting technique.

S'more perfection begins with a long, sturdy stick. You want to make sure you have the reach for getting the marshmallow to the optimum spot in the fire, which is not over the flames but near them. Remember, you are toasting, not burning. Find moderate, consistent heat and slowly rotate the marshmallow so that it toasts evenly, achieving that crispy golden brown state.

Waste no time once you remove the gooey confection from the fire. Promptly smash it between

chocolate bars and graham crackers. However, there are many other delectable options beyond these tried and true ingredients. Consider:

- Using Reese's Peanut Butter Cups instead of a plain chocolate bar
- Pressing the marshmallow between Pepperidge Farm Milano cookies
- Adding sliced strawberries, along with the marshmallow and dark chocolate, pressed between shortbread cookies
- Really uping the gourmet factor by adding peach slices and brie, along with dark chocolate and marshmallow pressed between graham crackers
- Using chocolate chip or peanut butter cookies instead of graham crackers
- Turning the dessert into a nightcap by dipping the marshmallows in Bailey's or other liquor

Resources

Cookware: For cast iron Dutch ovens and a wide variety of pots and pans, Lodge is the best. www.lodgemfg.com

For anodized aluminum pots, pans, and kettles, Four Dog Stove Company has a wide selection of quality cookware. www.fourdog.com

Grills/grates: The Grill On The Go made by Adjust-A-Grill is handy for cooking over coals or open flame. The device has a sturdy shelf that can move up and down over the fire and it is attached to single iron stake so hardware does not get in the way of the fire. www.adjustagrill.com

FIRE KEEPER: STEPHEN PYNE

During his prolific career as an author, Stephen Pyne has written more than 20 books on fire and become one of the most renowned authorities in the world on fire history. But he says he stumbled upon his life-long obsession simply by chance.

Right after graduating high school in Phoenix, Arizona, Pyne wandered up to the North Rim of Grand Canyon National Park in the summer of 1967 looking for a job. He was assigned to a wildland fire fighting crew, a job he would continue for 15 seasons as he completed undergraduate and graduate degrees in college. "Everything I've written, even the fact I write at all, dates from those years on the Rim," says Pyne.

Pyne studied history in college and he decided to apply those skills to researching the history and human cultural connection to fire. "During the 1970s there were no books about fire," says Pyne. "You had entire academic fields dedicated to other natural elements like water and earth sciences, but fire was not something that was studied. It was just supposed to be put out."

As Pyne dug through anthropological archives for his books, he discovered the pivotal role that fire played in human evolution and how it led to the progression toward modern civilization. Not only did early humans use fire for cooking and staying warm, but they also used it to herd wild game for hunting and burn fields for planting crops. Then the ash was used as fertilizer. Pyne argues that it was the use of fire that led to the development of agriculture. It also led to the invention of many technological innovations that further propelled humans up the food chain. "If fish and venison, maize and cassava could be cooked, why could fire not 'cook' the landscape for other goods as well?" writes Pyne in his book *Fire: A Brief History*. "Fire could break apart, distill, soften, stiffen, encrust, melt, or transmute. The range of things heated, steamed, boiled, or roasted is huge."

Cooking the landscape through perfecting the use of fire is what Pyne says led to the modern diet and the improved nutrition that nurtured brain development. In addition to burning fields as a seasonal part of agriculture for the staple crops of wheat, corn, and rice, pastures were also burned to support grazing animals raised for meat and dairy products.

In the early days of his career, Pyne drove around the West and lived out of a camper, writing about fire in places where the smoke was rising. In 1985, he began teaching at Arizona State University, where he has remained on the faculty. After five decades of writing about fire, Pyne has not only become a preeminent fire historian but also a leading authority on how fire might shape our future in an age of climate change.

In 2015 Pyne delivered a TED Talk in which he explained the intimate relationship between human evolution and the Earth's natural process of fire.

"We are a uniquely fire creature on a uniquely fire planet," he told the audience. "We are a keeper of the flame."

Embracing the Eternal Flame

*Creating a future with more natural fire
and less industrial combustion*

As someone who loves the wilderness and made a career out of writing about camping and hiking, I have sat around more campfires than I can count. The moments blur together—so many fires and so many happy memories of holding my son in my lap while he dozed off with a mug of hot chocolate in his hand. Or bonding with new friends as we told funny stories. Or sitting with my dog next to the glowing coals after everyone else had gone to bed.

But not all my outdoor experiences have been positive ones. And when I did encounter stressful or dangerous situations, the one constant was that I sought to build a fire. Like a steadfast friend, I knew I could count on fire to make things better.

When my son was in fourth grade, we joined another family for a backpacking trip in southern Utah's Grand Gulch Primitive Area. It was May and unseasonably hot as we hiked across steep and exposed stretches of slickrock domes and canyons. My son was only carrying water and a giant Harry Potter book in his daypack. I was carrying about 50 pounds of everything else. The other family offered to take some of my weight, so I gave them our sleeping bags and tent. They hiked ahead of us and somehow we got separated in an area where there was no marked trail.

My son and I took a break under a tree—the only shade around—and then I scouted the area to look for the trail or any sign of the people who had all our gear. No luck. We were lost and we had no way to stay warm once night fell in the high desert where the temperature would likely drop 40 degrees. Rather than wander around trying to find our trip mates, I decided we would stay put. We did not have shelter, sleeping bags, or food, but I had a lighter. I would build a fire.

In the fading twilight while my son read his Harry Potter book, I calmed myself by gathering a large amount of firewood. I sorted it into piles according to size and felt much better knowing that there was enough to keep a fire burning all night. We might not be very comfortable but we would not experience hypothermia.

Just as I was about to light the fire, two members of the other family who had been searching for us appeared. My son and I abandoned the giant pile of firewood and followed them to a campsite where all our modern gear waited.

After that, I always made sure that whenever I hiked in the wilderness I carried several methods of starting a fire. The near-survival situation in Grand Gulch taught me to stay put when lost, to build a fire, and plug in to the calm and comfort that only flames can provide.

For most of human history our use of fire has been as simple as that: It met our basic needs of survival. But in the last century things have changed. We have created what author Stephen Pyne describes as "a planet with too much combustion and too little fire."

In an ironic twist of fate, the very ingenuity that led human ancestors to harness fire a million years ago and enabled one species to rise above all others has now gone too far. Human-caused climate change from burning fossil fuels is warming the planet and altering the precious balance of atmospheric oxygen that first made combustion possible 120–200 million years ago.

In an age of climate change what should the role of fire be? Should we steer away from the burning of wood as well as fossil fuels?

Andy Stevenson, a silviculturist for the Coconino National Forest in northern Arizona, believes wood-burning fires should be part of the solution to climate change and not part of the problem.

"Wood heat could be a piece of the puzzle," says Stevenson. "If you are using a modern, high-efficiency stove that does not pollute the air and getting your wood from a sustainably managed forest, then you are heating your home in a carbon neutral way. Compared to burning fossil fuels, wood is a renewable energy resource that does not put extra CO_2 into the atmosphere."

As for the bigger environmental picture, Stevenson would like to see the general public in the United States embrace the role that fire plays in

maintaining forest health. In order to prevent cata-strophic, stand-replacing wildfires, which contribute to global warming, forest managers are increasingly conducting smaller-scale prescribed burns on public lands that have become overgrown after a century of fire suppression. This often means smoke temporar-ily settles over towns and annoys residents.

"Fire has been stigmatized for the last 150 years as bad," says Stevenson. "But fire is good hygiene for the forest. It cleans up the forest floor to guard against wildfires and the ashes provide essential nutrients to the soil." Stevenson is involved in the Four Forest Restoration Project, a $25 million initia-tive in northern Arizona using mechanical thinning and prescribed burns to return 2.4 million acres of national forest to ecological health. The ambitious decade-long program is aimed at restoring balance to ponderosa pine forest ecosystems and allow-ing regular cycles of low-intensity fire to return as nature intended.

To get back to a healthier relationship with fire, we need to remember our ancient connection to it. On an arduous backpacking trip in a remote corner of the Grand Canyon, I found myself, once again, in desperate need of a fire. I was with two other hiking partners and we had struck out from camp under blue skies on a long day hike. The weather suddenly turned and we ended up hiking eight miles back to camp in pouring rain and hail with no rain jackets. When we reached our tents after nightfall we were soaked to the bone, shivering and hungry. And it was still raining buckets.

We scrambled up to a sheltered alcove above our campsite, set up our stoves to cook dinner, and built a small fire to dry our wet clothes. Soon we were warm, laughing, and staring into the flames that cast dancing shadows on the alcove wall behind us. An experience that was miserable an hour earlier had become magical.

The next morning as we broke camp I walked back up to the alcove to make sure I didn't leave anything behind. As I surveyed the ground where we had been sitting the night before I saw that it was littered with ancient pottery shards. And prehistoric Native American pictographs of spirals, human fig-ures, and bighorn sheep decorated the alcove wall. Campers had been gathering around a fire in this spot for at least a thousand years. I lingered there for a moment relishing the connection to the land and the people who had also enjoyed this place long before me.

Fire reminds us of our roots. We are creatures of the flame.

Chimney Safety Institute of America: For a director of certified chimney sweeps and wood stove installation technicians, go to: www.csia.org.

U.S. Environmental Protection Agency: For information on wood stove and fireplace regulations as well as a list of manufacturers who meet EPA certification guidelines, go to: www.epa.gov/burnwise.

The Wood Heat Organization: This public outreach program of the Canadian government provides encyclopedic information on all aspects of home heating with wood in North America. Go to: www.woodheat.org.

COOKBOOKS

150 Best Recipes for Cooking in Foil: Ovens, BBQ, Camping, by Marilyn Haugen. Robert Rose Publishing, 2016.

Cooking with Fire: From Roasting on a Spit to Baking in a Tannur, Rediscovered Techniques and Recipes That Capture the Flavors of Wood-Fired Cooking, by Paula Marcoux. Storey Publishing, 2014.

NOLS Cookery, by Claudia Pearson. Stackpole Books, 1997.

Smoke: New Firewood Cooking, by Tim Byres. Rizzoli, 2013.

BIBLIOGRAPHY

Brown, Tom, Jr. *Tom Brown's Field Guide to Wilderness Survival.* Berkley: Berkley Books, 1983.

Burton, Frances. *Fire: The Spark that Ignited Human Evolution.* Albuquerque: University of New Mexico Press, 2009.

Cornell, Al. "The Role of Fire in the Domestication of Man," *Bulletin of Primitive Technology* no. 38, 74–76 (Fall 2009).

Frazer, James. *Myths of the Origin of Fire.* London: MacMillan, 1930.

Hough, Walter. "Fire as an Agent in Human Culture," *Smithsonian Bulletin 139* (1926).

Kochanski, Mors. *Bush Craft: Outdoor Skills and Wilderness Survival.* Edmonton, AB: Lone Pine Publishing, 2014.

Pyne, Stephen. *Fire: A Brief History.* Seattle: University of Washington Press, 2001.

Vvian, John. *Wood Heat.* Emmaus, PA: Rodale Press, 1976.

Wrangham, Richard. *Catching Fire: How Cooking Made Us Human.* New York: Basic Books, 2009.

INDEX

Italics indicate illustrations.